SpringerBriefs in Environmental Science

More information about this series at http://www.springer.com/series/8868

Naga Raju Maddela • Narasimha Golla
Rangaswamy Vengatampalli

Soil Enzymes

Influence of Sugar Industry Effluents
on Soil Enzyme Activities

 Springer

Naga Raju Maddela
Department of Life Sciences
Universidad Estatal Amazonica
Puyo, Pastaza, Ecuador

School of Environmental Science
  and Engineering
Sun Yat-Sen University
Xiaoguwei Island, Panyu District
Guangzhou, P.R. China

Rangaswamy Vengatampalli
Department of Microbiology
Sri Krishnadevaraya University
Anantapur, AP, India

Narasimha Golla
Department of Virology
Sri Venkateswara University
Tirupati, AP, India

ISSN 2191-5547   ISSN 2191-5555   (electronic)
SpringerBriefs in Environmental Science
ISBN 978-3-319-42654-9   ISBN 978-3-319-42655-6   (eBook)
DOI 10.1007/978-3-319-42655-6

Library of Congress Control Number: 2016951860

Printed on acid-free paper

This Springer imprint is published by Springer Nature
The registered company is Springer International Publishing AG
The registered company address is: Gewerbestrasse 11, 6330 Cham, Switzerland

Preface

Soil enzymes are key elements in the transformation of elements in the soil. Soil enzymes principally come from living and dead microbes, plant roots and residues, and soil animals. These enzymes are usually free from viable cells and stabilized in the soil matrix, where they accumulate or form complexes with organic matter (humus), clay, and humus–clay complexes. It is thought that two-thirds of enzyme activities can come from stabilized enzymes. Therefore, activity does not necessarily correlate with microbial number in a particular soil sample. This clearly indicates that enzyme activity is the cumulative effect of different biological systems present in the soil. However, although all members of the soil biota respond relatively to soil pollution, microbial communities are considered to be the first and swiftest responders to such environmental pollutants; because of their high sensitivity to respond to environmental changes, they play a fundamental role in the dynamics of organic matter and in the fragmentation of soils at different scales of time and space. On the other hand, certain enzymes reflect the activity of viable cells and occur in viable cells and not in stabilized soil complexes. Soil enzymes respond to soil management practices and act as good indicators of soil quality. They play an important role in organic matter decomposition and nutrient cycling. Some enzymes only break down organic matter, whereas others are involved in nutrient mineralization. Organic amendment applications, crop rotation, and cover crops have been shown to enhance soil enzyme activity. Soil enzymes are measured indirectly by determining their activity in a laboratory using biochemical assays.

This current book gives an overview of the impact of sugar industry effluents on selected soil enzyme activities. All the chapters were written by experts in the field, and our goal is that this book serves those who are interested in knowing soil enzyme activities under the influence of sugar industry effluents and how enzyme activities differ from soil enzyme activities affected by other effluents. Within the book, we have tried to address all aspects involved in this field: collection of soil, soil processing, physicochemical and biological characteristics of soil, soil incubation studies, soil enzyme assays, and the influence of sugar industry effluents on selected soil enzyme activities. The first four chapters focus on the general aspects of soil enzymes, and the last four chapters highlight the effluent's impact on soil enzymes. Altogether, the eight chapters describe the contents of this field precisely and clearly. In our view, this book provides exceptional information on soil enzyme activities in sugar industry polluted soils.

Puyo, Ecuador   Naga Raju Maddela, PhD
Tirupati, AP, India   Narasimha Golla, PhD
Anantapur, AP, India   Rangaswamy Vengatampalli, PhD

Contents

Chapter 1
Soil Collection

Soils are discontinuous heterogeneous environments that contain large numbers of diverse microbial populations including bacteria, fungi, algae, protozoa, and viruses. These populations vary with depth and soil type. In general, surface soil horizons have more organisms than subsurface horizons. Thus populations are influenced by many factors such as soil depth, soil type, and natural microsite variations. Natural microsite variations can allow very different microorganisms to coexist side by side in the same region of soil. Because of the great variability in soil microorganisms, it is always necessary to consider more than one sample during a microbial analysis of a site. Otherwise, it is not possible to get the complete picture of a selected soil. Thus, the sampling strategy is influenced by the goal of the analyses, the resources available, the site characteristics, and the history of the soil. The most accurate approach is multiple and individual analysis (MIA), which means taking many samples within a given site and performing a separate analysis of each sample. Another approach is composite analysis. An advantages of this approach is reducing time and effort by combining the multiple samples taken to form a composite sample; this in turn limits the number of analyses that must be performed. Thus composite sampling is better than the MIA approach. Another approach often used is to sample a site sequentially over time from a small defined location to determine effects on microbes. Such effects change over time; thus the effects are temporal.

Soil samples are usually collected using different tools or equipment as needed. For instance, bulk soil samples are easily obtained with a shovel (Fig. 1.1) or, better yet, a soil auger (Fig. 1.2). Soil augers are more precise than simple shovels because they ensure that samples are taken from exactly the same depth on each occasion. This is important, as several soil factors can vary considerably with depth, such as oxygen, moisture, and organic carbon content and soil temperature. Thus, soil augers are useful in characterizing the soils on the basis of depth. A simple hand auger is useful for taking shallow (up to depths of 6 ft) soil samples from areas that are unsaturated.

© Springer International Publishing Switzerland 2017
N.R. Maddela et al., *Soil Enzymes*, SpringerBriefs in Environmental Science,
DOI 10.1007/978-3-319-42655-6_1

Fig. 1.1 Soil shovel

Fig. 1.2 Soil auger

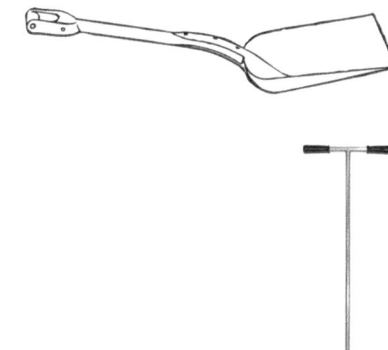

However, when samples are collected for microbial analysis, there is a possibility of sample contamination as the auger is pushed into the soil. Once the auger is inserted into the soil, microbes usually stick to the sides of the auger. When the auger pushes downward, it may contaminate the bottom part of the core. This causes erroneous results in the microbial profile of different soil depths. However, this problem can be overcome by using a sterile spatula to scrape away the outer layer of the core and using the inner part of the core for further analysis. Furthermore, the sample collected in this manner may not be truly representative of the site. This is due to the varied nature of soil and limited diameter of an auger. Thus, it is always better to collect several samples and prepare a composite sample. This greatly reduces the total number of samples and associated costs of the analyses that are performed. Proper procedures should be followed while preparing the composite samples. The foremost requisite for the preparation of a composite sample is the selection of a wide and uniform area. Then equal amounts of soil samples are collected and placed in a sterile bucket or plastic bags. Later, these samples are mixed and become the composite sample. Another advantage of a composite soil sample is that if the sample seems big enough for storage, a portion of the composite sample can be removed, and leftover sample can be analyzed. In order to get the precise data of the soil, samples should be stored on ice until further processed and analyzed. In other incidences, the experimental aim is to test the effect of a soil amendment (such as fertilizer, pesticide, or sewage sludge) on microbial populations over untreated control. In such a case, a sample of each treatment must be analyzed separately and compared with the untreated control. There is also another method of soil collection, called randomized sampling. It involves choosing points randomly within the selected site.

Soil samples can also be collected in different directions of a selected site. For instance, samples are collected in a single direction called transect sampling. This type of sampling is very useful if a sampling site is situated on the bank of a river called a riparian area (Fig. 1.3). Transects could be chosen adjacent to a streambed and at right angles to the streambed. This type of sampling greatly helps in the

Fig. 1.3 Riparian area

evaluation of influence of a stream on microbial populations. There is another method called two-stage sampling, which is suitable when a site consists of a hillside slope and a level plain. During sampling by this method, the area is broken into regular subunits called primary units. Furthermore, subsamples can be taken randomly or systematically within each primary unit. On the other hand, grid sampling is used when little is known about the variability within the soil of a mapping area. In this type of sampling, samples are taken systematically at regular intervals at a fixed spacing.

In the present investigation, soil samples (test) were collected from different sites, where effluents are being discharged by Sri. Rayalaseema Sugars and Energy (Pvt.) Limited (Nandyal Sugar Factory), Ayyalurimetta village, and Nandyal, Kurnool district of Andhra Pradesh. A soil sample without sugar mill effluents (control) was collected from a site adjacent to the sugar mill. These two soil samples were air dried and mixed thoroughly to increase homogeneity and shifted to <2 mm sieves for determination of soil texture.

Chapter 2
Soil Physicochemical Properties

Introduction

Soil is a complex matter and comprises minerals, soil organic matter, water, and air. These fractions greatly influence soil texture, structure, and porosity. These properties subsequently affect air and water movement in the soil layers, and thus the soil's ability to function. Therefore, soil physicochemical properties have a great influence on the soil quality. Soil texture especially can have a profound effect on many other properties. Thus, soil texture is considered one of the most important physical properties of soil. In fact, soil texture is a complex fraction, consisting of three mineral particles, such as sand, silt, and clay. These particles vary by size and make up the fine mineral fraction. Generally, the coarse mineral fraction, which consists of particles over 2 mm in diameter, is not considered in texture. But in some cases they may affect soil physicochemical properties such as water retention. The textural category of a soil is decided by the relative amount of various particles sizes in a soil, that is, whether it is clay, loam, sandy loam, or another (Fig. 2.1).

Usually the soil texture is a result of a "weathering" process, which is the physical and chemical breakdown of rocks and minerals, inasmuch as soil is a heterogeneous substance in terms of composition and structure. Therefore, these fractions weather at different rates. This in turn affects a soil's texture. For instance, easily weathered rock forms clay-rich soils. On the other hand, granite is a slow-weathering rock that forms sandy and coarse soils. Once soil texture is formed, it is not altered easily by management practices; instead it is fairly constant because weathering is a relatively slow process and is not easily subject to changes.

Soil structure has a complex arrangement. Soil particles are bound together into larger clusters, called aggregates or "peds." This aggregation plays an important role in many soil parameters. It increases stability against erosion, maintains porosity and soil water movement, and improves soil fertility and carbon sequestration in the soil (Nichols et al. 2004). For instance, if spheroidal

© Springer International Publishing Switzerland 2017   5
N.R. Maddela et al., *Soil Enzymes*, SpringerBriefs in Environmental Science,
DOI 10.1007/978-3-319-42655-6_2

Fig. 2.1 Soil texture

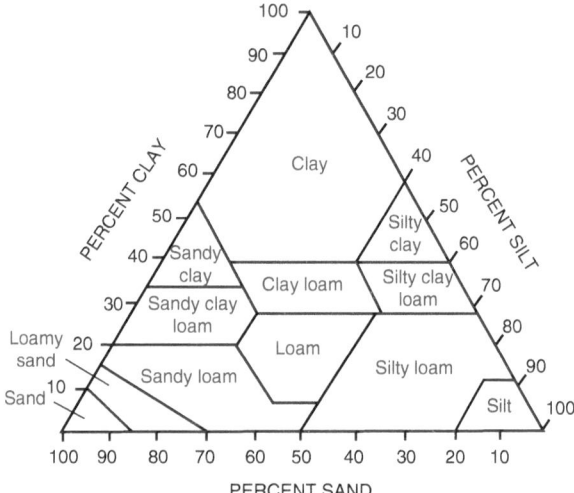

peds are loosely packed and glued together by organic substances, it gives a granular structure to the soil. A granular structure frequently appears in A horizons where there is a large amount of soil organic matter and much biological activity. On the other hand, B horizons usually contain larger peds, which are arranged in the form of plates, blocks, or prisms. This type of formation involves shrink/swell processes and adhesive substances (Gardiner and Miller 2004). Generally, soil swells as it becomes wet or freezes, and shrinks when it dries and thaws. During soil swelling and shrinking processes, cracks form around soil masses, creating peds. These peds are held together with the help of many substances such as organic materials, iron oxides, clay, and carbonates. Transportation of water, air, solutes, and deep water drainage through the soil is usually mediated by cracks and channels. Mechanically, finer soils are much stronger with a more defined structure than coarser soils. This is because shrink/swell processes predominate in clay-rich soils and there is more cohesive strength between particles.

Colloid surfaces are good platforms for the most chemical interactions that occur in the soil. This is primarily because of their chemical make-up, and large surface area. Due to their charged surfaces, colloids are able to sorb, or attract ions with the soil solution. The attracted molecules can be adsorbed on the colloid surface or exchanged with other ions and released into the soil solution. The adsorption or exchange of molecules is mainly dependent on the ion's charge, size, and concentration in the soil. The ability of a soil to sorb and exchange ions is called its "exchange capacity". In general, both positive and negative charges are present on colloid surfaces. If soils are dominated by negative charges, they will have an overall (net) negative charge and vice versa. Furthermore, negatively charged soils will have more attraction towards cations to exchange sites than anions. Such soils tend to have greater cation exchange capacities (CEC) than anion exchange capacities

(AEC). This is opposite in the case of positively charged soil. Generally, fine-structured soils have a greater exchange capacity than coarse soils because of higher proportions of colloids.

In view of microbiological activity in the soil, another important soil physico-chemical property is the soil's pH. It refers to a soil's acidity or alkalinity and is the measure of hydrogen ions (H^+) in the soil. A high amount of H^+ corresponds to a low pH value and vice versa. The surface charge of colloids is greatly influenced by soil pH; this in turn affects the CEC and AEC. Whenever there is a higher concentration of H^+ (lower pH), the negative charge on the colloids will be neutralized, thereby decreasing CEC and increasing AEC. The opposite occurs when pH increases. Such types of changes in the soil's pH will have the greatest impact on soil microbial activities, because microbial growth is pH dependent.

Inorganic salts present in the soil play a vital role in affecting the soil properties. It is a well-known fact that the presence and concentration of salts in soil can have adverse effects on soil function and management. Salt-rich soils are most common in arid and semiarid regions because in these regions evaporation exceeds precipitation and dissolved salts are left behind to accumulate. Also, higher levels of salts are seen in soils where vegetation or irrigation changes have caused salts to leach and accumulate in low-lying places (saline seeps). Based on the concentration and type of salts, there are three main types of salt-affected soils: saline, sodic, and saline-sodic. Saline soils contain a high amount of soluble salts, primarily calcium (Ca^{2+}), magnesium (Mg^{2+}), and potassium (K^+), whereas sodic soils are dominated by sodium (Na^+). Saline-sodic soils have both high salt and Na^+ content. Salts in soil can affect structure, porosity, and plant/water relations that can ultimately lead to decreased productivity.

Another considerable chemical complex in the soil is carbonates of calcium or magnesium. Such soils are referred to as "calcareous" soils. Weathering of carbonate-rich parent material, such as limestone or lime-enriched glacial till, results in the formation of calcareous soils. Generally it occurs in areas where precipitation is too low to leach the minerals from the soil. The location of carbonates varies in the calcareous soils. They can be found throughout a soil profile or concentrated in the lower horizons due to downward leaching. The calcareous horizon layer is denoted as the "k" layer. The effervescence (fizz) reaction that occurs when a drop of dilute acid (10% hydrochloric acid or strong vinegar) is applied is the test for distinguishing calcareous soils (Brady and Weil 2002).

In fact, soil's carbonates have the greatest impact on several soil physicochemical characteristics. For instance, carbonates can affect soil productivity by influencing soil pH, structure, WHC, and water flow. However, calcareous soils have a high "buffering capacity", or resistance to changes in pH. This is due to free carbonates being able to neutralize acids in the soil effectively. Thus, the pH of calcareous soils changes very little and alkaline conditions (near pH 8) are maintained. Carbonates can also alter soil structure by affecting texture and promoting aggregation. WHC can be affected by the size and concentration of carbonates. Very fine carbonate particles can coat clay and silt particles and reduce their surface tension with water, and when a large percentage of $CaCO_3$ is present in the clay fraction (30% or higher), the soil's WHC can be reduced (Massoud 1972). Nevertheless, high concentrations of carbonates can be toxic to plants and soil organisms.

Another important soil chemical is gypsum ($CaSO_4 \cdot 2H_2O$), however, it is predominantly found in semi-arid soils. Its accumulation process in the soil is very similar to that of carbonate accumulation. However, gypsum is more soluble than carbonates and sulfates and is not as abundant as carbonate. Gypsum deposits are less common and generally found in drier climates where very little leaching occurs (Lindsay 1979). Soils dominated by gypsum are buffered, but not to the extent of carbonate-dominated soils, and typically have a pH between 7 and 8.

Organic amendments, such as crop residues, animal manures, logging and wood manufacturing residues, various industrial organic wastes, sewage wastes, and food processing and fiber harvesting wastes, are naturally occurring compounds that are used as additives to improve soil physical conditions and/or plant nutrition (Donahue et al. 1983). Industrial organic residues include waste from processes such as alcohol distillation, paper making, meat packing, flour making, and petroleum refining. A wealth of information is available on industrial effluents and their influence on soils. Discharge of effluents from various industries such as a dying factory (Swaminathan and Ravi 1987), paper mills (Mishra and Sunandashaoo 1989; Singh et al. 2005), tannery and chromate industry (Nandakumar 1990), petrochemical industries (Andrade 2002), and cotton ginning mills (Narasimha et al. 1999) influenced the physicochemical properties of soil.

Discharge of soils by combined pulp and paper mill effluents increased soil pH, organic carbon, nitrogen, potassium, and phosphorus (Kannan and Oblisami 1990a). Application of sodium-based black liquor from fiber pulping for paper making increased soil pH and electrical conductivity (Xiao et al. 2005). Soils treated with distillery effluents showed very high electrical conductivity and potassium ions (Devarajan et al. 2002). Soils discharged with food waste compost showed increased pH, electrical conductivity, total nitrogen content, and organic matter (Kim et al. 2002). Coir dust of coir fiber industries increased the water-holding capacity of soil (Devi et al. 2002). Disposal of sewage sludge on agricultural land can, however, promote changes of soil physical properties (Smith 1991). Continuous irrigation of soils with paper mill effluents showed changes in pH, electrical conductivity, and organic carbon (Chinnaiah et al. 2002).

An increase in municipal organic compost in soil significantly increased the organic matter percentage, nitrogen, available phosphorus, and exchangeable potassium over the control treatment in all treated levels (Chuasavathi and Trelo-ges 2001). Similarly, there was an increase in the pH (Renukaprasanna et al. 2002; Adhikari et al. 1994), water-holding capacity, and electrical conductivity (Renukaprasanna et al. 2002) of sewage-irrigated soils. Long-term (20 years) municipal waste disposal on soil increased soil organic matter, pH, and total nitrogen (Anikew 2002).

In contrast, application of sewage effluents decreased the soil pH and increased electrical conductivity (Bhogal et al. 2002). Andrade (2002) reported that the continuity of the porous space of the soil matrix is impeded by the presence of pollutants, which generate areas highly limiting to water flow. Urbanization has numerous impacts on soils. Gilbert (1991) and Craul (1992) summarized differences in physical properties of urban and nonurban soils, including reduced soil structure, compaction, surface crusting, restricted aeration and drainage, and modified temperature regime.

Impact of Effluents

The mineral matter of soil samples such as sand, silt, and clay content was analyzed using different sizes of sieves following the method of Alexander (1961). Hundred percent water-holding capacity of soil samples was measured by finding the amount of the water added to both soil samples to get the saturation point. The 60% water-holding capacity of soil samples was calculated by the method of Johnson and Ulrich (1960). Soil pH was measured at 1:1.25 soil-to-water ratio in the Elico digital pH meter with calomel glass electrode assembly. The organic carbon content in both soil samples was determined by the Walky and Blaky method (Jackson 1971); electrical conductivity of soil samples with/without effluent was determined by the addition of 100 ml distilled to 1 g of soil sample in an Elico conductivity meter. Total nitrogen content in the both soil samples was determined by the method of Microkjeldhal (Jackson 1971).

The soil sample discharged with sugar industry effluents underwent changes in all measured parameters of physical and chemical properties in comparison to the control sample (Table 2.1). The soil textures in terms of percentage of sand, silt, and clay were continuously 51, 19, 30 and 64, 22, and 14 in test and control, respectively. The above results indicated that effluent-discharged soil had relatively lower sand, silt, and higher clay contents than control soil. Similarly, long-term application to soil of sewage effluents (Abdelnainm et al. 1987) and cotton ginning mill effluents (Narasimha et al. 1999) led to an increase in clay content and improved soil texture and structure.

Table 2.1 Physicochemical properties of soil as affected by sugar industry effluents

Properties	Controla	Testb
Color	Gray	Thick black
Odor	Normal	Bad
pH	8.30	7.62
Texture:		
Sand (%)	64	51
Silt (%)	22	29
Clay (%)	14	20
Electrical conductivity (μMhos/cm)	0.24	1.71
Water-holding capacity (ml/g)	0.28	0.34
Organic matter (%)	3.602	6.432
Total nitrogen (g/Kg)	0.14	0.22
Available potassium (K_2O) in Kg/A	170	332
Available phosphorus (P_2O_5) in Kg/A	1.5	12.0
Calcium	Low	High

[a]Control: Soil without sugar industry effluents.
[b]Test: Soil polluted with sugar industry effluents.

Higher water-holding capacity and electrical conductivity were observed in contaminated soil than in control; values were continuously 0.34 ml/g, 1.71 mMhos/cm and 0.28 ml/g, 0.24 mMhos/cm, respectively (Table 2.1). Increased water-holding capacity and electrical conductivity in contaminated soil might be due to accumulation of organic wastes and salts in the sugar industry effluents. Similarly, soil discharged with effluents from cotton ginning mills (Narasimha et al. 1999) and paper mills (Medhi et al. 2005) increased the water-holding capacity and electrical conductivity. In contrast, soils polluted by cement industries had low water-holding capacity and higher electrical conductivity (Shanthi 1993; Sivakumar and De Brito 1995). The pH of polluted soil was reduced from 8.30 to 7.62 upon release of sugar industry effluents (Table 2.1). Slightly lowered pH in polluted soil than the control soil can be explained in terms of release of effluents that were acidic in nature, containing agro-based chemicals from industry. Zende (1995) reported that discharge of sugarcane residues from industry reduced soil pH. Organic matter content of contaminated soil was 6.432 g/kg and control soil 3.602 g/kg. Higher organic matter content of polluted soil may be due to the discharge of effluents in an organic nature. The contents of total nitrogen and phosphorous in effluent soil were continuously 0.22 g/kg, 8.21 mg/g against 0.14 g/kg and 4.25 mg/g of control. Similar reports were made by Narasimha et al. (1999): discharge of effluents from the cotton ginning industry increased the total nitrogen and phosphorous content compared to the control soil.

Chapter 3
Soil Microbiological Properties

Introduction

Microorganisms are the smallest living systems, but their activities are numerous. They represent the largest and most diverse biotic group in soil. Fertile soil usually contains 10^6–10^9 bacteria per gram of soil (Tugel and Lewandowski 1999). They play a vital role in soil texture by their organic secretions. Microorganisms are usually higher in the A horizon, and they contribute to the formation of the granular structure. Soil microorganisms consist of both prokaryotes and eukaryotes, including bacteria, protozoa, algae, fungi, and actinomycetes. However, in terms of bio-geo-chemical cycles, soil bacteria are very important. They are crucial in soil organic matter (SOM) decomposition, nutrient transformations, and small clay aggregation. Certain bacteria carry out unique roles in the soil. For instance, Rhizobia are nitrogen-fixing bacteria associated with legume roots, whereas protozoans (e.g., amoebas, ciliates, flagellates) are mobile microorganisms and live as heterotrophes by feeding on other soil microorganisms, and SOM. Algae are autotrophes like plants, photosynthesize, and are found near the soil surface. Another important soil microbial community is fungi, a diverse group of microorganisms. They are primarily responsible for the breakdown of SOM and have large aggregate stability. Structurally they have long hyphae or mycelia; these can spread yards to miles underneath the soil surface. These filaments help bind soil particles. Another considerably important soil microbial community is actinomycetes. These microorganisms have features of bacteria and fungi. They are prokaryotes like bacteria and have filamentous structures like fungi. They play a vital role in the degradation of SOM containing more resistant fractions. Also, actinomycetes are the principal agents in giving an earthy odor to soil. Nevertheless, bacteria dominate in agricultural and grassland soils, whereas fungi are more prevalent in forest and acidic soils (Tugel and Lewandowski 1999).

Mycorrhizae, mutualistic associations existing between plants and fungi, are another important aspect of soil biology and are found in almost all soils and plants.

© Springer International Publishing Switzerland 2017
N.R. Maddela et al., *Soil Enzymes*, SpringerBriefs in Environmental Science,
DOI 10.1007/978-3-319-42655-6_3

In this relationship, fungi infect a plant and live in, or on, its root. In this symbiotic association, the fungus depends on the plant for energy and, in turn, the fungus and its hyphae can take up nutrients for the plants. These associations have been shown to increase plant–water relations and reduce the severity of some plant diseases (Smith and Read 1997), as well as improve soil aggregate stability. This is due to the binking actions of hyphae and glomalin, a mychorrhizally secreted chemical (Nichols et al. 2004). But the beneficial effects of mycorrhizae to a plant depend on the soil conditions and the plant's requirements. Currently, there are many commercial mycorrhizal inoculants available; additional research is necessary to understand the effectiveness of mycorrhizal inoculants towards plant growth (Smith and Read 1997). Nevertheless, there are some possibilities for encouraging the mycorrhizal symbioses in agriculture; these include improving and maintaining existing mycorrhizal populations by increasing SOM content, reducing tillage and other soil disturbances, and eliminating long fallow periods.

Biological Activity

Soil biological activity is very complex and controlled by many different factors, for instance, residue and quantity and quality of SOM, in which nitrogen content is the principal limiting factor for soil organism activity. However, other factors cannot be ignored, for example, oxygen, pH, temperature, and moisture. In general, highest soil biological activities are being observed in soils with adequate levels of oxygen, near-neutral pH, 85–95°F temperature, and 50–60% moisture (Brady and Weil 2002). Nevertheless, combinations of the above factors are essential for the optimum levels of soil biological activities. There are certain extremophiles that have been adapted to extreme environmental conditions, but generally speaking, overall activity diminishes when conditions fall outside these ideal ranges. For instance, oxygen diffusion is impeded if a soil becomes too wet. This causes overall soil biological activity to diminish because oxygen is required by most soil organisms.

On the other hand, soil management practices can greatly influence soil biological activities because such practices are responsible for the changes in aeration and structure, cropping systems, and different amendments of nutrients and xenobiotics. Nevertheless, the effects of soil management practices vary from type to type. For example, tillage typically accelerates the bacteria and protozoa activities in the soil for a short period because tillage increases the aeration and breaks up residue into smaller particles that are more exposed to microbial attack (Vigil and Sparks 2003). Fungal biomass has been shown to increase conservation tillage systems; this may be due to less tillage disrupting fungal hyphal networks and/or increases in SOM levels (Frey et al. 1999). Usually, soil management practices that do increase SOM levels and minimize soil disturbances will help to increase the earthworm populations in the soil, and crop rotation systems usually support more microbial diversity and activity than monoculture systems. This may be due to increased and more diverse residues plus specific interactions occurring between certain plants and organisms (Olfert et al. 2002).

Nutrient amendments (fertilization) can also influence the quantity and activity of soil organism populations. In general, fertilizer addition to the soils containing low levels of nutrients or SOM causes increased biological activity. Activities will be higher if fertilizers contain N, and populations will eventually stabilize as N is consumed. However, certain fertilizer applications may cause temporary harm to soil biological activity. For instance, injection of anhydrous ammonia can reduce the soil organisms temporarily at the injection site (Tugel and Lewandowski 1999). But later most organism populations will rebound with time.

Regarding the microbial analyses of soil samples, it is always preferable to perform the analyses as soon as possible after the collection of a soil. Therefore, effects of storage on microbial populations can be greatly reduced. Generally, once the sample is collected from the field, changes in microbial populations will not depend on the method of storage. Microbial numbers and their activities were reduced even when soil samples were stored in a field moist condition at 4 °C for only 3 months (Stotzky et al. 1962). However, although bacterial populations were changed, actinomycete populations remained unchanged. The first and foremost step in soil microbial analysis usually involves sieving through a 2-mm mesh. This process removes the large stones and other debris from the surface soil sample. However, most often the soil sample is air dried to facilitate easy sieving. It should be noted that the soil moisture content should not become low because low moisture content can reduce the microbial populations in the soil sample (Sparkling and Cheshire 1979). After the sieving, if samples needs to be stored, 4 °C is preferable. However, during short-term storage, two things should be remembered: one is drying and another is development of anaerobic conditions. Both of these can greatly alter microbial populations. Generally, storage up to 21 days appears to leave most soil microbial properties unchanged (Wollum 1994), however, time is very important with respect to microbial analysis. Most often, sampling of surface soils does not require sterile procedure. Inasmuch as these soils are continually exposed to the local atmosphere, it is believed that such exposure during sampling and processing will not affect the results significantly. But more precautions are needed if subsurface soils are to be processed. Generally, subsurface samples have lower cultural counts. Therefore, outside microbial contaminants may significantly affect the number that actually present. Also, subsurface sediments are not frequently exposed to the atmosphere. Therefore microbial contaminants present in the surface atmosphere may pollute the subsurface sediments if not properly handled or processed. Additionally, collection of subsurface samples is always more expensive than the collection of surface samples; often there is no second chance at collection for subsurface samples. More often subsurface samples are obtained by coring, and may be either immediately frozen and sent back to the analysis site as an intact core or processed at the coring site itself. But in either case, the core surface normally should be scraped off by sterile spatula or a subcore can be collected by a small-diameter plastic syringe. Such sample is then placed in a sterile plastic bag and analyzed immediately or kept at low temperature (freezing conditions) for future analysis.

Traditional soil microbial analysis methods have usually involved either cultural or direct count assays. Culturing assays are done by dilution and plating methodology on selective and differential media. Direct counts give information about the total number of bacteria present but don't tell us about the viable, nonviable, and diversity of populations present. Plate counts enumerate the total culturable or selected cultural populations, therefore they provide information on the different populations present. However, it is well known that often less than 1% of soil bacteria is culturable (Amann et al. 1995). Therefore, cultural information offers only a small fraction of the actual count. Additionally, the medium selected for cultural counts will select the populations that grow best on the particular medium. Thus, selection of the medium is crucial in determining the microbial populations of the selected soil sample.

In the recent past, a powerful technique was developed to obtain and study total DNA extracted from soil bacteria. This technique allows us to know how many different kinds of bacteria are present in the selected soil sample as well as their genetic potential. However, this technique has its own limitations. Therefore researchers more often now use DNA extraction in conjunction with direct and cultural counts in order to obtain the data with accuracy. Generally, bacterial DNA from soil samples can be obtained in two ways: fractionation of bacteria from soil followed by cell lysis and DNA extraction, and in situ lysis of bacteria within the soil matrix and released DNA extracted subsequently.

Impact of Effluents

The microbial population in soil is an indication of soil fertility. Microbial activity in soil is strongly influenced by minerals, colloids, and humates that bind organic chemicals, inorganic ions, and water films to surfaces (Bollag et al. 2002). Microbial enzymes are involved in complex relationships with other components of the soil system. The efficiency of intracellular enzymes depends on the conditions that microorganisms face. As a result of the activity of extracellular enzymes in microorganisms, they are able to begin functioning on suitable substrates in effluents discharged into the soils. Populations of fungi and bacteria in the rhizosphere of food waste compost in the soil significantly increased soil carbon, enzyme activities, and microbial populations (Kim et al. 2002). Treatment of soils with sodium-based black liquor from fiber pulping for paper making (Xiao et al. 2005), effluents of pulp and paper mill (Kannan and Oblisami 1990a), alcohol industry (Monanmani et al. 1990), and cotton ginning mills (Narasimha et al. 1999) increased the soil microbial biomass. In contrast, urban soils have less numbers and diversity of organisms compared with natural or seminatural soils (Gilbert 1991; Harris 1991).

Microflora of soil such as bacterial population in the soils contaminated with/ without effluents of the sugar industry were enumerated by taking 1 g of soil and serially diluting it up to 10^{-6}–10^{-9}. Suspension of 0.1 ml was platted on the nutrient agar medium and spread with a sterilized spreader. Plates were incubated at 37 °C for 24 h. After 24 h, the bacterial colonies appearing on the surface of the medium were counted by colony counter.

Quantity of bacteria was expressed in terms of colony-forming units/g soil by using the formula:

$$\text{Colony forming units} / \text{g soil} = \text{Number of colonies} / \left[(\text{Sample volume in mL}) \times (\text{Dilution factor}) \right]$$

The fungal population in soil samples with/without effluents of the sugar mill was isolated and enumerated on Martin's rose bengal medium by the serial dilution technique of 1 g of soil in sterilized distilled water. The soil suspension was spread on the medium and incubated for 7 days at 28 °C. After the incubation, fungal colonies appeared on the medium, calculated based on the color and morphology. The total number of fungi was enumerated according to the method described previously for bacteria, and the quantity was expressed in terms of cfu/g of soil.

The microflora of both soil samples were enumerated and are listed in Table 3.1. Threefold higher bacterial and twofold higher fungal populations were observed in the test sample over control. The higher microbial population in sugar industry effluent-discharged soil could probably be due to the presence of high organic matter and low pH in the soil due to release of acidic effluents rich in organic matter content. Similar to these results, soils polluted with effluents of alcohol (Monanmani et al. 1990) and cotton ginning mills (Narasimha et al. 1999) also showed increased microbial populations.

Table 3.1 Biological properties[a] of soil as affected by sugar industry effluents

Microorganism	Control[b]	Test[c]
Bacteria	64×10^4	192×10^4
Fungi	7×10^4	15×10^4

[a]Microbial population in terms of colony forming units g^{-1} of soil
[b]Soil without sugar industry effluents
[c]Soil polluted with sugar industry effluents

Chapter 4
Soil Incubation Studies

Soil quality cannot be measured directly because it is a broad, integrative, context-dependent concept. Instead, we analyze a variety of proxy measurements that together provide clues about how a soil is functioning as viewed from one or more soil-use perspectives. These measurements are called soil quality indicators. A set of low-cost, readily measured indicators that accurately predict soil functions of interest is called an efficient indicator set. Indicators of soil quality may include characteristics of soil solids, soil solutions, soil atmospheres, vegetation, and other soil biota, and possibly even economic analyses of land-use or ecosystem services.

Although the quantity and quality of data may differ, the process of soil quality evaluation follows the same basic steps regardless of the method used: identification of soil use issues followed by indicator selection and interpretation. More specifically, to select appropriate indicators, one must first determine the land-use objectives, and then indicators must be proposed, measured, and assessed across a representative set of land and management practices. An efficient indicator set should be used to inform land management decisions at specific sites and then be used to monitor trends in soil function after changing practices and over time.

For the determination of soil enzyme activities such as protease, cellulase, amylase, and invertase, 5 g of soil sample was placed in each test tube (25×200 mm) and 60 % water-holding capacity was maintained with the addition of the required amount of distilled water into soil and tubes were kept in an incubator at 28 ± 4 °C by replacing water during incubation. Triplicate soil samples with/without effluent discharges were drawn after 0, 10, 20, 30, and 40 days of incubation to determine soil enzyme activities.

© Springer International Publishing Switzerland 2017   17
N.R. Maddela et al., *Soil Enzymes*, SpringerBriefs in Environmental Science,
DOI 10.1007/978-3-319-42655-6_4

Chapter 5
Soil Protease

Introduction

Proteases are widely distributed among soils and show a wide range of activities (Ladd and Butler 1972; Hayano 1986). Protease enzymes are involved in the initial hydrolysis of protein components of organic nitrogen to simple amino acids. Hydrolytic degradation of proteins is an important step in the nitrogen cycle. Proteases in soils hydrolyze not only added proteins but also native soil added proteins (Dedeken and Voets 1965). Protease enzymes, detected in microorganisms, plants, and animals, catalyze the hydrolysis of proteins to polypeptides and oligopeptides to amino acids (Handa et al. 2000) involved in the nitrogen cycle (Moreno et al. 2003). Treatment of soils with metal-contaminated sewage sludge (Achberger and Ohlinger 1988), effluents from cotton ginning mills (Narasimha 1997), and pig slurry (Plaza et al. 2002) increased protease activity. In contrast to this, decreased protease activity was observed in soils treated with herbicides (Pahwa and Bajaj 1999), insecticides (Omar and Abd-Alla 2000), organic matter (Ladd and Butler 1969), crude oils (Walker et al. 1975), and chlorothalonil (Singh et al. 2002).

Enzyme Assay

Triplicate samples of soils with/without effluent discharges were incubated in the manner specified in Chap. 4; soil samples were withdrawn at desired intervals (0, 10, 20, 30, and 40 days) to determine protease activity by the method described by Speir and Ross (1975). Five grams of soil samples were shifted to test tubes (25×200 mm); to this, 10 ml of 2 % casein in 0.1 M tris buffer at pH 7.5 were added and incubated for 24 h. Another set of soil samples treated in the same manner with modification of casein was replaced by this buffer without substrate. All the tubes were incubated for the desired incubation periods. After incubation 4 ml of 17.5 % trichloroacetic

© Springer International Publishing Switzerland 2017   19
N.R. Maddela et al., *Soil Enzymes*, SpringerBriefs in Environmental Science,
DOI 10.1007/978-3-319-42655-6_5

acid was added to these samples and the suspension was filtered by Whatman No. 1 filter paper. The amount of protein in the filtrate was determined by the Folin–Lowery method (Lowry et al. 1951) using a Elico digital spectrophotometer. Finally, the protease activity was expressed in terms of milligrams of tyrosine equivalents per gram of soil. Similarly, another three sets of control soil samples were treated with 10, 50, and 100 % effluents, respectively, and protease activities were assessed.

Impact of Effluents

Soil enzyme protease is excreted by soil microorganisms, plants, and animals by means of their metabolic activities. Protease is an extracellular enzyme secreted by soil microorganisms and is distributed among soils exhibiting a wide range of activities (Ladd and Butler 1972).

At room temperature, the protease activity of soil samples discharged with/without sugar industrial effluents was measured with the supplementation 2 % caseinate, with 60 % water-holding capacity and activity was measured in terms of tyrosine equivalents formed in trichloroacetic acid soluble fraction during 24 h, and the results are listed in Table 5.1. By increasing the soil incubation days, the protease activity was also increased up to a 30-days interval, then activity declined in both polluted and nonpolluted soil. Maximum activity was recorded in soil samples discharged with effluents rather than control with improvement from two- to threefold in test soil. For instance, at 0 day interval, soil samples with effluent discharges exhibited 0.284 mg of tyrosine equivalent g^{-1} (mg TE g^{-1}) against 0.069 mg TE g^{-1} of control soil. This trend followed at all incubation intervals ranging from two- to threefold (Table 5.1). Increased proteolytic activity in soils with effluent discharges may be due to high availability of substrates (casein) and increased proteolytic microorganisms in effluent soil. A similar report was made by others, such as soils treated with the effluents of a cotton ginning mill (Narasimha 1997), tomato processing waste (Sarade and Richard 1994), and pig slurry (Plaza et al. 2002) improved soil protease activity more than control soil. In contrast, soils polluted with cement dust from cement industries (Shanthi 1993) and wastewater treatment plant discharge (Montuelle and Volat 1998) ceased the protease activity. Similarly, in soils treated with fungicides (Sreenivasulu 2005), protease activity was increased by increasing the incubation, maximum at 20 days.

Protease activity in soils without amendment of substrate (casein) was studied and results are listed in Table 5.2. By increasing the incubation period in both soil samples the protease activity accelerated up to 30 days interval and then declined in both soil samples. For instance, at 0 day, the test sample showed 0.042 mg TE g^{-1}; this increased by 5166 % at 30 days, and later it decreased by 14 % at 40 days interval. Compared to the two soil samples, protease activity was maximum in effluent discharged soil than in control soil (Table 5.2) at all incubations. Positive protease activity in soils without supplementation of substrates such as casein may be due to

Table 5.1 Protease activity[a] in soil (with substrate) after 24 h[b] incubation as influenced by sugar industrial effluents

Incubation (in days)	Control[c]	Test[d]
0	0.069 (100)	0.284 (411)
10	0.093 (100)	0.669 (719)
20	0.845 (100)	2.569 (307)
30	1.622 (100)	4.156 (256)
40	1.300 (100)	3.970 (305)

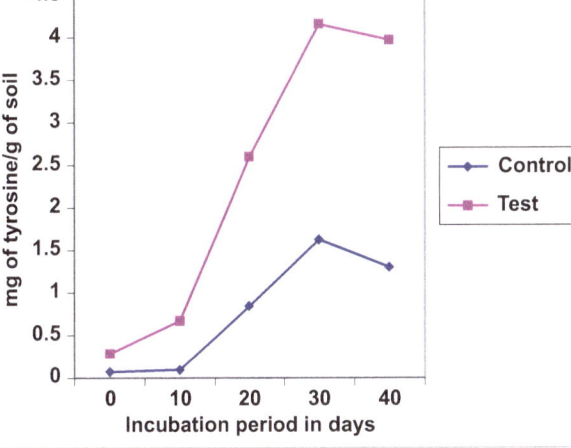

Figures in parentheses indicate relative production percentages
[a]Milligrams of tyrosine g^{-1} of soil
[b]Incubation, in hours, of soil with casein (2 % w/w)
[c]Soil without sugar industrial effluents
[d]Soil polluted with sugar industrial effluents

the presence of protein and related compounds in the soil. Microorganisms inhabiting soil are proteolytic and can show protease activity even without supplementation of substrate (casein) in the soil. In a similar observation made by Narasimha (1997), soils discharged with/without effluents of a cotton ginning mill without supplementation of substrate improved protease activity in polluted soil more than in control soil.

Protease activity in soil samples with effluents in different concentrations including 10, 50, and 100 % were observed with the amendments of substrate (casein) and results are reported in Table 5.3. By increasing the soil incubation days, protease activity was also increased up to 30 days then declined in all concentrations of effluents (Table 5.3). For instance, the protease activity at 0 day interval in control was 0.069 mg TE g^{-1} and this activity was increased 2250 % at 30 days interval then declined 20 % at 40 days interval. The same trend was followed at 10, 50, and 100 % effluents (Table 5.3). Comparison of protease activity between control and different concentrations of effluent-treated soils, by increasing the effluent concentrations,

Table 5.2 Protease activity[a] in soil (without substrate) after 24 h[b] incubation as influenced by sugar industrial effluents

Incubation (in days)	Control[c]	Test[d]
0	0.034 (100)	0.042 (123)
10	0.044 (100)	0.240 (545)
20	0.248 (100)	1.192 (481)
30	0.724 (100)	2.212 (305)
40	0.348 (100)	1.892 (544)

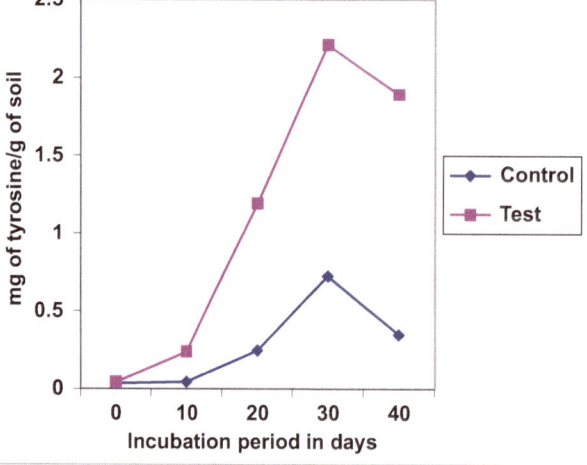

Figures in parentheses indicate relative production percentages
[a]Milligrams of tyrosine g^{-1} of soil
[b]Incubation, in hours, of soil without casein (2 % w/w)
[c]Soil without sugar industrial effluents
[d]Soil polluted with sugar industrial effluents

showed the protease activity was also increased, with maximum at 100%, at all incubations. For instance, at 0 day interval, control soil showed 0.069 mg TE g^{-1}, whereas 10, 50, and 100% samples, showed 0.204, 0.358, and 0.38 mg TE g^{-1}, respectively. The same results were observed at the rest of the incubations. Similar results were reported by Plaza et al. (2002): treatment of soil with pig slurry resulted in higher protease activity observed at higher concentrations of this residue. In contrast, Sreenivasulu (2005) reported that, at higher concentrations of fungicide in soil, protease activity was decreased.

The protease activity in soil samples treated with different concentrations of effluents, without addition of substrate was studied and the results are depicted in Table 5.4. Here also, similar results were obtained; by increasing the effluent concentration, the activity of protease was also increased, to a maximum of 100%, at all incubations. For example, at 0 day, the 100% sample showed maximum activity,

Table 5.3 Protease activity[a] in soil (with substrate) after 24 h[b] incubation as influenced by different concentrations of sugar industrial effluents

Incubation (in days)	Different concentrations of effluents (%)			
	0	10	50	100
0	0.069 (100)	0.204 (296)	0.358 (519)	0.380 (362)
10	0.093 (100)	0.393 (422)	0.573 (616)	0.69 (742)
20	0.845 (100)	1.134 (134)	1.422 (168)	1.664 (197)
30	1.622 (100)	2.208 (136)	2.311 (142)	2.578 (159)
40	1.300 (100)	2.022 (155)	2.20 (169)	2.494 (192)

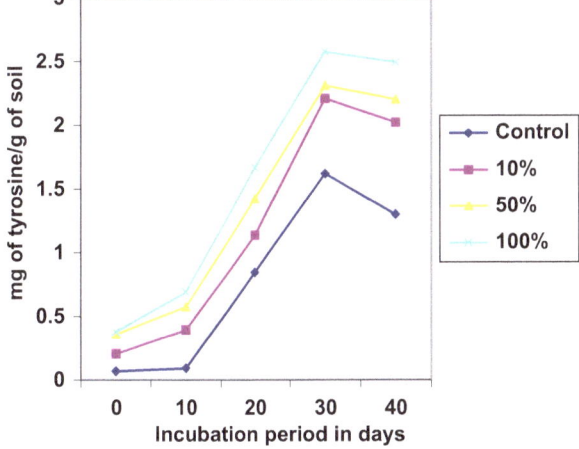

Figures in parentheses indicate relative production percentages
[a]Milligrams of tyrosine g^{-1} of soil
[b]Incubation, in hours, of soil with casein (2 % w/w)

that is, 0.016 mg TE g^{-1}, whereas control, 10, and 50 % samples showed 0.034, 0.01, and 0.018 mg TE g^{-1} activities, respectively. In each sample, with increasing the incubation period from 0 to 30 days, the activity increased, but further it dropped at 40 days. For instance, the control sample exhibited 0.034 mg TE g^{-1} activity at 0 day; it increased to 0.724 mg TE g^{-1}, and later it decreased to 0.348 mg TE g^{-1} at 40 days. A similar trend was followed in the remaining three samples of different concentrations of effluents.

Table 5.4 Protease activity[a] in soil (without substrate) after 24 h[b] incubation as influenced by different concentrations of sugar industrial effluents

Incubation (in days)	Different concentrations of effluents (%)			
	0	10	50	100
0	0.034 (100)	0.010 (29)	0.018 (53)	0.016 (47)
10	0.044 (100)	0.034 (78)	0.068 (154)	0.844 (1918)
20	0.248 (100)	0.688 (277)	0.852 (343)	1.240 (500)
30	0.724 (100)	0.96 (132)	1.436 (198)	1.484 (205)
40	0.348 (100)	0.732 (210)	1.248 (359)	1.316 (378)

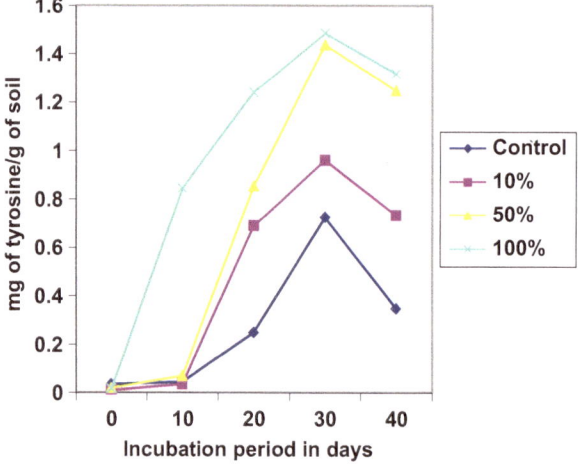

Figures in parentheses indicate relative production percentages
[a]Milligrams of tyrosine g^{-1} of soil
[b]Incubation, in hours, of soil without casein (2 % w/w)

Chapter 6
Soil Cellulase

Introduction

Cellulase is a core enzyme; it consists of exo-, endo-, and β-glucosidases. This enzyme synergistically acts on cellulose polymer substrates, which are abundantly available on the earth's surface in the form of wood, chips, rocks, and municipal waste. Cellulose is the most abundant polysaccharide in plant cell walls and represents a significant input to soils (Richards 1987). Cellulose hydrolysis into glucose is mainly achieved by the complex enzyme cellulase produced by fungi (Maile and Linkins 1978). These enzymes are extensively studied in plant litter (Wood and Bat 1988; Sinsabaugh and Linkins 1987; Linkins et al. 1990). Liberation of extracellular enzymes of cellulase by microbes during litter decomposition may be influenced by many factors including temperature, moisture, pH, and substrate concentration (Linkins et al. 1984).

Cellulase activity is indicated by degradation of substrates such as the cellulose polymer of cellophane (Markus 1955; Kiss and Peterfi 1959), cellulose powder (Rawald et al. 1968), and carboxy methyl cellulose (Kong and Dommergues 1972), and its activity was measured by Pancholy and Rice (1973) through the appearance of reducing sugars measured spectrophotometrically. Cellulase activity was potentially correlated with fungal and bacterial population in soil (Joshi et al. 1993).

Little information is available on the effect of sugar industrial effluents on soil cellulase activity. Cellulase activity was enhanced in soils irrigated with effluents of the textile and sugar industries (Kannan and Oblisami 1990b), cotton ginning mills (Narasimha 1997), solid urban waste (Ramakrishna Parama et al. 2002), sodium-based black liquor from fiber pulping for paper making (Xiao et al. 2005), and paper mill effluents and amendment additions (Chinnaiah et al. 2002). Urban expansion into wild lands significantly increases cellulase activity (Douglas and Oleksyszyn 2002). By contrast, soil contaminated with cement dust from cement industries (Shanthi 1993) and crude oils (Walker et al. 1975) ceased cellulase activity in soils.

© Springer International Publishing Switzerland 2017   25
N.R. Maddela et al., *Soil Enzymes*, SpringerBriefs in Environmental Science,
DOI 10.1007/978-3-319-42655-6_6

Enzyme Assay

Triplicate samples of soils with/without effluent discharges were incubated in the manner specified in Chap. 4; soil samples were withdrawn at desired intervals (0, 10, 20, 30, and 40 days) to determine cellulase activity by the method described by Pancholy and Rice (1973). Five grams of soil samples were placed in 50-ml Erlenmayer flasks and 0.5 ml of toluene was added. All the contents in the flasks were mixed thoroughly and after 15 min, 10 ml of acetate buffer at pH 5.9 were added, followed by 1 % carboxy methyl cellulose (CMC). Another set of soil samples was treated in the same manner by replacing CMC with buffer without substrate. Then flasks were incubated for 30 min and approximately 50 ml of distilled water were added. The suspension was filtered by Whatman No. 1 filter paper and the volume of the filtrate was made up to 100 ml with distilled water. The amount of the reducing sugar content in the filtrate was determined by the Nelson–Somagyi method (1944) in an Elico digital spectrophotometer. Similarly, another three sets of control soil samples were treated with 10, 50, and 100 % effluents, respectively, and cellulase activities were assessed.

Impact of Effluents

Cellulase plays an important role as a group of enzymes in global recycling of the most abundant polymer, cellulose in nature. The impact of sugar industrial effluents on cellulase activity has been studied in polluted and nonpolluted soil samples supplemented with substrate (1 % carboxy methyl cellulose), described previously and the results are listed in Table 6.1. The results showed that the cellulase activity in the test sample was higher than the control sample at all incubations. The increase of cellulase activity of the test sample range was between 22 and 57 % over control. By increasing the incubation interval, the activity of cellulase was also increased in both test and control samples, at a maximum interval of 30 days. For instance, in the test sample, at 0 day, the cellulase activity was 36.66 µg glucose equivalents g^{-1} against 23.33 µg GE g^{-1} of control; it was increased by 117 % at 30 days interval, then declined by 30 % at 40 days interval in the test sample. Increased cellulase activity in soils with effluent discharges may be due to high availability of substrates (CMC) and increased cellulolytic microorganisms in effluent soils. Similar results were reported by others: discharge of effluents from paper mill and pressmud addition (Chinnaiah et al. 2002), cotton ginning mill (Narasimha 1997), potassium-based black liquor from straw pulping (Xiao et al. 2005), urban waste (Ramakrishna Parama et al. 2002), and tomato processing waste (Sarade and Richard 1994) increased the cellulase activity in soil compared to control. Similarly, by increasing the incubation period, cellulase activity in with/without fungicide-treated soils increased at a maximum interval of 20 days; later the activities decreased (Sreenivasulu 2005). According to Joshi et al. (1993), cellulase activity was greatly

Table 6.1 Cellulase activity[a] in soil (with substrate) after 30 min[b] incubation as influenced by sugar industrial effluents

Incubation (in days)	Control[c]	Test[d]
0	23.33 (100)	36.66 (157)
10	43.33 (100)	63.33 (146)
20	63.33 (100)	80.00 (126)
30	90.0 (100)	110.0 (122)
40	55.0 (100)	76.66 (139)

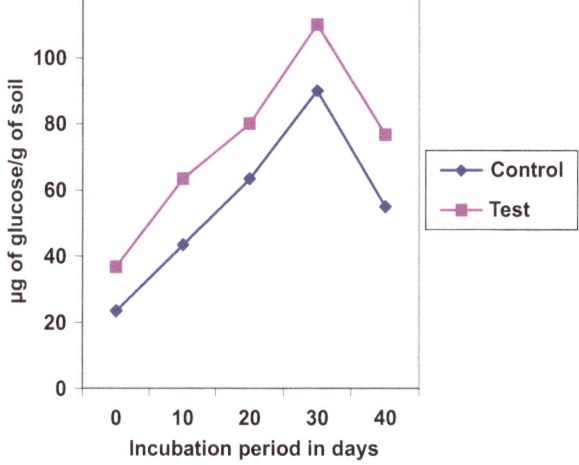

Figures in parentheses indicate relative production percentages
[a]Micrograms of glucose g[-1] of soil
[b]Incubation, in minutes, of soil with carboxy methyl cellulose (1 % w/w)
[c]Soil without sugar industrial effluents
[d]Soil polluted with sugar industrial effluents

increased in soils treated with cellulose and increased cellulase activity was positively correlated with fungal and bacterial number and moisture content of litter. A high significant correlation between cellulase activity and soil respiration was observed (Splading 1979) as well as microbial biomass (Kanazawa and Miyashita 1987 and Donnelly et al. 1990).

Cellulase activity has been studied in polluted and nonpolluted soil samples supplemented without substrate (1 % CMC); it has been described in the previous section and results given in Table 6.2. The results showed that the cellulase activity in the test sample was higher than the control sample at all incubations. The increase of cellulase activity of the test sample range was between 25 and 67 % over control. By increasing the incubation period, the activity of cellulase was increased in both test and control samples, at a maximum interval of 30 days. For instance, in the test sample, at 0 day, the cellulase activity was 16.66 μg GE g[-1] against 13.33 μg GE g[-1] of control; it was increased by 296 % at 30 days interval, then declined by 80 % at

Table 6.2 Cellulase activity[a] in soil (without substrate) after 30 min[b] incubation as influenced by sugar industrial effluents

Incubation (in days)	Control[c]	Test[d]
0	13.33 (100)	16.66 (125)
10	23.33 (100)	36.66 (157)
20	33.33 (100)	53.33 (160)
30	40.00 (100)	66.66 (167)
40	6.66 (100)	13.33 (200)

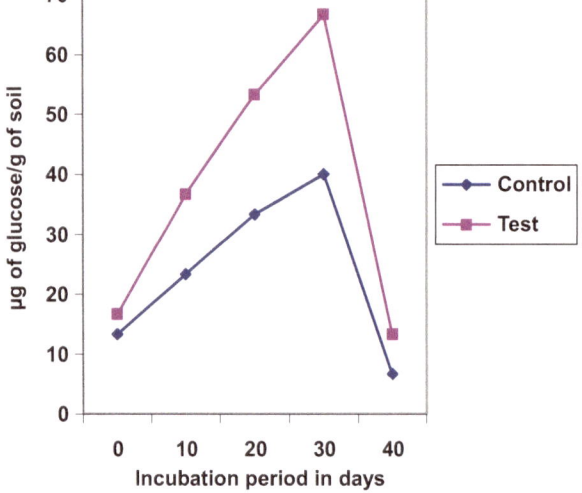

Figures in parentheses indicate relative production percentages
[a]Micrograms of glucose g^{-1} of soil
[b]Incubation, in minutes, of soil without carboxy methyl cellulose (1 % w/w)
[c]Soil without sugar industrial effluents
[d]Soil polluted with sugar industrial effluents

40 days interval in the test. A similar report made by Narasimha (1997): without amendment of substrate, discharges of effluents from a cotton ginning mill improved soil cellulase activity more than control soil.

Cellulase activity with supplementation of substrate in soil samples treated with various concentrations of effluents such as 10, 50, and 100 % were assessed, as described earlier, and results are reported in Table 6.3. According to the results, the sample with 50 % effluent showed higher activity at all incubations over control, 10, and 100 % samples. For instance, at 0 day, the 50 % sample showed 33.33 µg GE g^{-1} cellulase activity, whereas control, 10, and 100 % samples showed 23.33, 16.66, and 26.66 µg GE g^{-1} cellulase activities, respectively. At the same time, the 10 % sample has shown less activity and the 100 % sample has shown more activity over control at all incubations. In each sample, with increasing the incubation period, the cellulase activity increased at a maximum interval of 30 days. For instance, in

Table 6.3 Cellulase activity[a] in soil (with substrate) after 30 min[b] incubation as influenced by different concentrations of sugar industrial effluents

Incubation (in days)	Different concentrations of effluents (%)			
	0	10	50	100
0	23.33 (100)	16.66 (71)	33.33 (143)	26.66 (114)
10	43.33 (100)	40.00 (92)	70.00 (161)	46.66 (108)
20	63.33 (100)	43.33 (68)	106.66 (168)	76.66 (121)
30	90.0 (100)	83.33 (92)	126.66 (141)	96.66 (107)
40	55.0 (100)	76.66 (139)	123.33 (224)	74.0 (134)

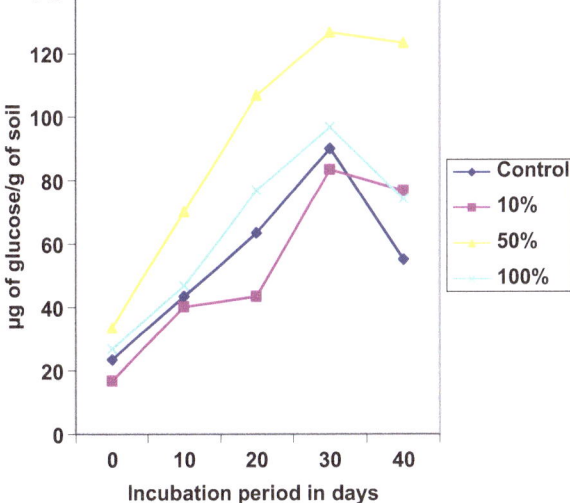

Figures in parentheses indicate relative production percentages
[a]Micrograms of glucose g^{-1} of soil
[b]Incubation, in minutes, of soil with carboxy methyl cellulose (1 % w/w)

the 10 % sample, at 0 days, the activity was 16.66 µg GE g^{-1}; it increased by 400 % at 30 days, and later it decreased by 8 % at 40 days. The same trend was followed in the remaining samples. Similar results were reported by Ramakrishna Parama et al. (2002): soil treated with urban waste along with additives such as cow dung, rock phosphate, green leaves, and coir dust increased the cellulase activity in the initial incubation and later it was stabilized. Decreased activity of cellulase at higher concentrations of effluent may be due to the exposure of cell free enzyme to highly concentrated effluent. This is correlated with the reports of Gaianfreda and Bollag (1994, 1996): soil organic matter may have an inhibitory effect on enzymatic activity in the terrestrial system. Inhibition of enzymatic activity at higher concentrations of effluents may also be due to high acidity (Ruggiero et al. 1996); this is also correlated with the pH of the polluted sample. Many enzymes are short lived in soil environments (Ahn et al. 2002). Similar observations were made by Sreenivasulu (2005): a high concentration of fungicide in soil inhibited soil cellulase activity.

Table 6.4 Cellulase activity[a] in soil (without substrate) after 30 min[b] incubation as influenced by different concentrations of sugar industrial effluents

| Incubation (in days) | Different concentrations of effluents (%) | | | |
	0	10	50	100
0	13.33 (100)	6.66 (50)	33.33 (250)	13.33 (100)
10	23.33 (100)	16.66 (71)	43.33 (186)	23.33 (100)
20	33.33 (100)	33.33 (100)	53.33 (160)	43.33 (130)
30	40.00 (100)	83.33 (208)	63.33 (158)	53.33 (133)
40	6.66 (100)	76.66 (1151)	26.66 (400)	19.66 (295)

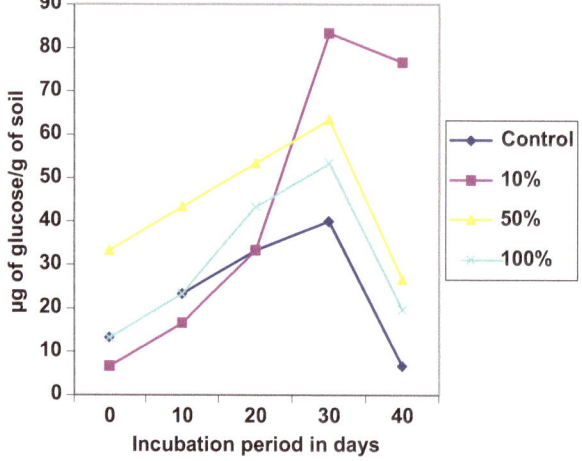

Figures in parentheses indicate relative production percentages
[a]Micrograms of glucose g^{-1} of soil
[b]Incubation, in minutes, of soil without carboxy methyl cellulose (1 % w/w)

Cellulase activity without supplementation of substrate in soil samples treated with various concentrations of effluents including 10, 50, and 100 % were assessed, and results reported in Table 6.4. According to the results, soil with 50 % effluent has shown higher cellulase activity at all incubations over the rest of the samples. For instance, at 0 day, the 50 % sample showed 33.33 µg GE g^{-1} cellulase activity, whereas control, 10, and 100 % samples showed 13.33, 6.66, and 13.33 µg GE g^{-1} cellulase activities, respectively. Similar results were observed in the rest of the incubations. In each sample, by increasing the incubation period, the cellulase activity was also increased at a maximum interval of 30 days. For instance, at 0 day, in the 10 % sample, the activity was 6.66 µg GE g^{-1}; it was increased by 1150 % at 30 days, and later decreased by 8 % at 40 days. The same trend was followed in the remaining three samples.

Chapter 7
Soil Amylase

Introduction

Amylases are widely distributed in soils and have a wide range of activities (Ladd and Butler 1972) and properties (Ladd and Butler 1969). Starch is a major carbon compound within most plant tissues and increases during active photosynthesis and decreases as it is enzymatically converted into sugars. Amylase catalyzes the hydrolytic depolymerization of polysaccharides in soil (Tu and Miles 1976). Starch-hydrolyzing enzymes are usually extracellular and inducible, but their activity depends on the type of substrate (Alexander 1977). Soil amylase is responsible for the major breakdown of complex polysaccharides including starch to a readily available form of glucose (Singaram and Kamalakumari 2000). Production of these extracellular enzymes from microbes during litter degradation may be influenced by temperature, moisture, pH, and substrate involvement (Linkins et al. 1984; Sinsabaugh and Linkins 1987). Amylase activity was significantly correlated with fungal and bacterial populations and moisture content of litter (Joshi et al. 1993). Changes in amylase activity during litter decomposition were attributed to changes in microbial populations (Ross and Roberts 1973). Increased amylase activity was observed when soil was treated with insecticides and pesticides (Tu 1982), effluents released from pulp and paper mills (Kannan and Oblisami 1990b), cotton ginning mills (Narasimha 1997), and pressmud plus paper mill effluents (Chinnaiah et al. 2002). By contrast, amylase activity was reduced when soil was treated with imidacloprid (Tu 1995), dimethoate (Mandic et al. 1997), and chlorothalonil (Singh et al. 2002).

© Springer International Publishing Switzerland 2017   31
N.R. Maddela et al., *Soil Enzymes*, SpringerBriefs in Environmental Science,
DOI 10.1007/978-3-319-42655-6_7

Enzyme Assay

Triplicate samples of soils with/without effluent discharges were incubated in the manner specified in Chap. 4: soil samples were withdrawn at desired intervals at 0, 10, 20, 30, and 40 days to determine amylase activity using the method developed by Cole (1977) and followed by Tu (1981a, b). Five grams of soil samples were placed in the test tubes (25 × 200 mm); to this 1 ml of toluene was added. All the contents in the tubes were mixed thoroughly; after 15 min, 6 ml of 2 % starch in 0.2 M acetate buffer (pH 5.5) was added. Another set of soil samples was treated in the same manner by replacing starch with acetate buffer without substrate. Tubes were incubated for 48 h. The suspension was filtered by Whatman No. 1 filter paper, and the amount of reducing sugar content in the filtrate was determined by the Nelson–Somagyi method (1944) using an Elico digital spectrophotometer. Similarly, another three sets of control soil samples were treated with 10, 50, and 100 % effluents, respectively, and amylase activity was assessed. With another set of soil samples, the amylase activities were assessed after a 72-h incubation period as mentioned above.

Impact of Effluents

The amylase enzyme plays a crucial role in catalyzing the hydrolysis and solubilization of starch. Starch-hydrolyzing enzymes are usually extracellular and inducible. The amylase activity in test and control soil samples was measured by incubating the samples for 48 h in the presence of substrate (2 % starch), as described previously; results are listed in Table 7.1. The test sample showed higher activity over control at all incubations; the increasing percentage of the test sample was between 25 and 133 over control. Both samples showed higher activity at 30 days interval and then activities declined. For instance, the test sample exhibited 620 µg GE g^{-1} amylase activity against 550 µg GE g^{-1} of control at 0 day interval; later it increased by 179 % at 30 days and declined by 35 % at 40 days interval in test. The increased amylase activity in polluted soil over control may be due to the availability of substrate and/or amylolytic microflora in polluted soil. Similar results were reported by others: the addition of pressmud and paper mill effluent irrigation (Chinnaiah et al. 2002), effluents from pulp and paper mill (Kannan and Oblisami 1990b), cotton ginning mill (Narasimha 1997), and fungicides (Sreenivasulu 2005) improved soil amylase activity. In contrast, soil polluted with cement dust from the cement industry ceased soil amylase activity (Shanthi 1993).

The soil amylase activity in test and control soil samples was measured by incubating for 48 h without supplementation of substrate, described in Chapter 4 and results are depicted in Table 7.2. The test sample exhibited higher activity over control at all incubations ranging from 33 to 60 %. Both samples have shown higher activities at 30 days interval; later activities were lower. For instance, the test sample exhibited 80 µg GE g^{-1} activity at 0 day against 60 µg GE g^{-1} of control. It increased by 170 % at 30 days and declined by 65 % at 40 days in test. Comparatively at all

Table 7.1 Amylase activity[a] in soil (with substrate) after 48 h[b] incubation as influenced by sugar industrial effluents

Incubation (in days)	Control[c]	Test[d]
0	550 (100)	620 (125)
10	610 (100)	1420 (233)
20	850 (100)	1500 (176)
30	900 (100)	1930 (214)
40	360 (100)	1250 (347)

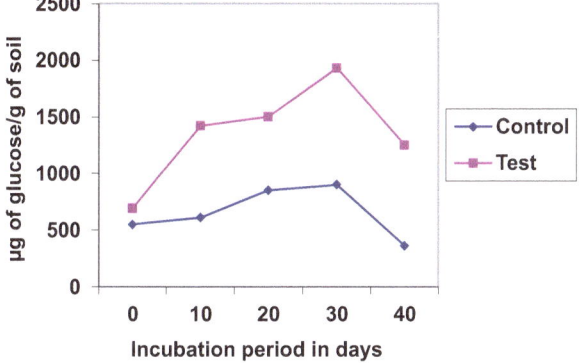

Figures in parentheses indicate relative production percentages
[a]Micrograms of glucose g^{-1} of soil
[b]Incubation, in hours, of soil with starch (2 % w/w)
[c]Soil without sugar industrial effluents
[d]Soil polluted with sugar industrial effluents

Table 7.2 Amylase activity[a] in soil (without substrate) after 48 h[b] incubation as influenced by sugar industrial effluents

Incubation (in days)	Control[c]	Test[d]
0	60 (100)	80 (133)
10	85 (100)	120 (141)
20	110 (100)	150 (136)
30	135 (100)	216 (160)
40	80 (100)	76 (95)

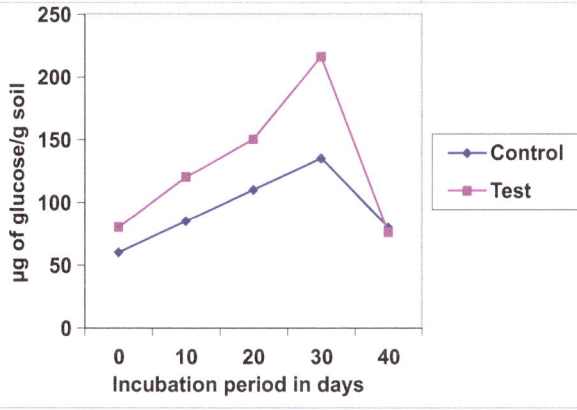

Figures in parentheses indicate relative production percentages
[a]Micrograms of glucose g^{-1} of soil
[b]Incubation, in hours, of soil without starch (2 % w/w)
[c]Soil without sugar industrial effluents
[d]Soil polluted with sugar industrial effluents

Table 7.3 Amylase activity[a] in soil (with substrate) after 48 h[b] incubation as influenced by different concentrations of sugar industrial effluents

Incubation (in days)	Different concentrations of effluents (%)			
	0	10	50	100
0	550 (100)	760 (138)	200 (36)	100 (18)
10	610 (100)	780 (128)	300 (49)	200 (33)
20	850 (100)	940 (110)	420 (49)	310 (36)
30	900 (100)	1130 (125)	750 (83)	550 (61)
40	360 (100)	230 (64)	245 (68)	150 (41)

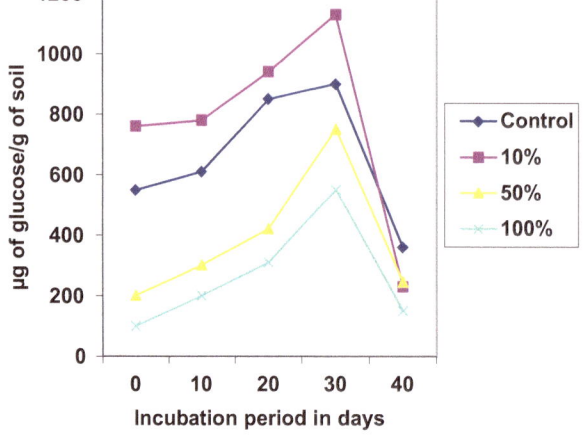

Figures in parentheses indicate relative production percentages
[a]Micrograms of glucose g^{-1} of soil
[b]Incubation, in minutes, of soil with starch (2 % w/w)

incubations, the amylase activity without substrate in test and control was lower with substrate.

Amylase activity in soils treated with different concentrations of effluents such as 10, 50, and 100 % was measured by incubating the samples for 48 h with the addition of substrate (2 % starch) as explained previously and the results are reported in Table 7.3. By increasing the concentration of effluents, the amylase activity increased up to 10 % effluent; thereafter it decreased at all incubations. For instance, at 0 day, 10 % effluent-treated soil exhibited 760 µg GE g^{-1}, whereas control, 50, and 100 % samples showed 550, 200, and 100 µg GE g^{-1}, respectively. The same trend was followed at all incubations. Moreover, 50 and 100 % samples showed less activity than control at all incubations. Similarly, in soils treated with fungicides (Sreenivasulu 2005) of different concentrations, amylase activity (24 h) was increased by increasing the concentrations of fungicide. By increasing the incubation period, amylase activity was also increased in all samples, with the maximum at 30 days; later it declined. For instance, the 10 % sample showed 760 µg GE g^{-1} activity at 0 day, it was increased by 48 % at 30 days, and then reduced by 80 % at

Table 7.4 Amylase activity[a] in soil (without substrate) after 48 h[b] incubation as influenced by different concentrations of sugar industrial effluents

Incubation (in days)	Different concentrations of effluents (%)			
	0	10	50	100
0	60 (100)	90 (150)	68 (113)	50 (83)
10	85 (100)	110 (129)	88 (103)	70 (82)
20	110 (100)	150 (136)	135 (123)	120 (109)
30	135 (100)	165 (122)	158 (117)	138 (102)
40	80 (100)	90 (112)	70 (87)	55 (69)

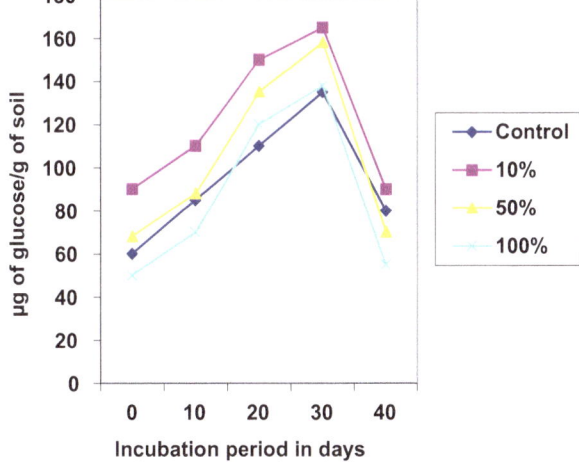

Figures in parentheses indicate relative production percentages
[a]Micrograms of glucose g^{-1} of soil
[b]Incubation, in minutes, of soil without starch (2 % w/w)

40 days interval. The same results were observed in the other three samples at all incubations.

Amylase activity in soils treated with various concentrations of effluents was measured by incubating the samples for 48 h without amendment of substrate, and results are reported in Table 7.4. Here also similar results were obtained. By increasing the concentration of effluent, the amylase activity increased up to 10 % effluent concentration; later it decreased at all incubations. For instance, at 0 day interval, the 10 % effluent-treated sample showed 90 μg GE g^{-1} activity, but control, 50, and 100 % samples exhibited 60, 68, and 50 μg GE g^{-1} activity, respectively. At the same time, by increasing the incubation period, the amylase activity increased, with maximum at 30 days; later it decreased in all concentrations. For instance, the 10 % sample showed 90 μg GE g^{1} at 0 day, it increased by 83 % at 30 days, and then declined by 45 % at 40 days interval. The overall amylase activity without substrate was comparatively less at all incubations and concentrations with substrate.

Table 7.5 Amylase activity[a] in soil (with substrate) after 72 h[b] incubation as influenced by sugar industrial effluents

Incubation (in days)	Control[c]	Test[d]
0	180 (100)	380 (211)
10	780 (100)	840 (108)
20	1000 (100)	1180 (118)
30	1300 (100)	1740 (134)
40	820 (100)	950 (116)

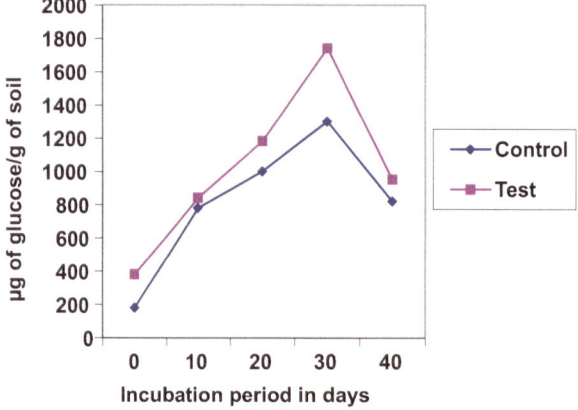

Figures in parentheses indicate relative production percentages
[a]Micrograms of glucose g^{-1} of soil
[b]Incubation, in hours, of soil with starch (2 % w/w)
[c]Soil without sugar industrial effluents
[d]Soil polluted with sugar industrial effluents

The activity of amylase in test and control soil samples was measured by incubating for 72 h in the presence of substrate (2 % starch), and the results are listed in Table 7.5. The test sample showed higher activity over control at all incubations; the increasing percentage of the test sample was between 8 and 111 over control. Both samples showed higher activities at 30 days interval and then activities declined. For instance, the test sample exhibited 380 μg GE g^{-1} amylase activity against 180 μg GE g^{-1} of control at 0 day interval; later it increased by 358 % at 30 days and declined by 45 % at 40 days interval in test. The increased amylase activity in polluted soil over control may be due to availability of substrate and more amylolytic microflora in the polluted soil. Comparatively, polluted soils showed less amylase activity at 72 h than 24 h.

The soil amylase activity in test and control soil samples were measured by incubating for 72 h without supplementation of substrate, and the results are depicted in Table 7.6. The test sample exhibited higher activity over control at all incubations, and the increased percentage of the test sample ranged from 9 to 100. Both samples

Table 7.6 Amylase activity[a] in soil (without substrate) after 72 h[b] incubation as influenced by sugar industrial effluents

Incubation (in days)	Control[c]	Test[d]
0	20 (100)	40 (200)
10	70 (100)	90 (128)
20	92 (100)	100 (109)
30	110 (100)	121 (110)
40	25 (100)	40 (160)

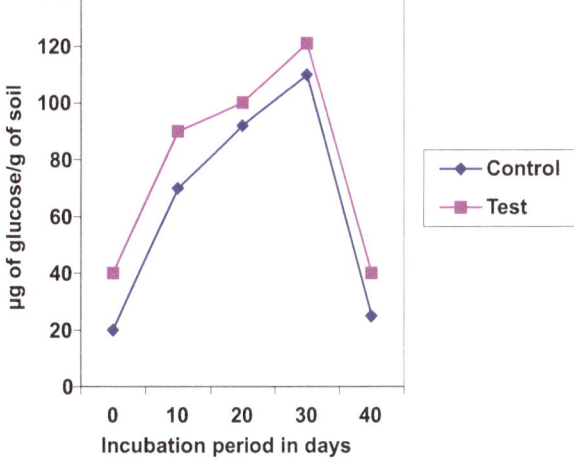

Figures in parentheses indicate relative production percentages
[a]Micrograms of glucose g^{-1} of soil
[b]Incubation, in hours, of soil without starch (2 % w/w)
[c]Soil without sugar industrial effluents
[d]Soil polluted with sugar industrial effluents

showed higher activities at 30 days interval; later activities were lower. For instance, the test sample exhibited 40 µg GE g^{-1} activity at 0 day against 20 µg GE g^{-1} of control. It was increased by 202 % at 30 days and declined by 67 % at 40 days in test. Comparatively at all incubations, the amylase activity without substrate in test and control were less than with substrate.

Amylase activities in soils treated with different concentrations of effluents such as 10, 50, and 100 % were measured by incubating the samples for 72 h with addition of substrate (2 % starch), and results are reported in Table 7.7. The soil sample with 50 % effluent showed slightly higher activities of amylase over control, whereas the 10 and 100 % samples showed less activity than control at all incubations. For example, at 0 day, the 50 % sample showed 800 µg GE g^{-1} activity, whereas control, 10, and 100 % samples showed 180, 500, and 600 µg GE g^{-1} activity, respectively. Similar results were reported by Sreenivasulu (2005) when soils treated with different concentrations of fungicides by increasing the concentration of fungicide, the

Table 7.7 Amylase activity[a] in soil (with substrate) after 72 h[b] incubation as influenced by different concentrations of sugar industrial effluents

	Different concentrations of effluents (%)			
Incubation (in days)	0	10	50	100
0	180 (100)	500 (278)	800 (444)	600 (333)
10	780 (100)	720 (92)	920 (118)	620 (79)
20	1000 (100)	900 (90)	980 (98)	810 (81)
30	1300 (100)	990 (76)	1110 (85)	908 (70)
40	820 (100)	700 (85)	900 (110)	800 (97)

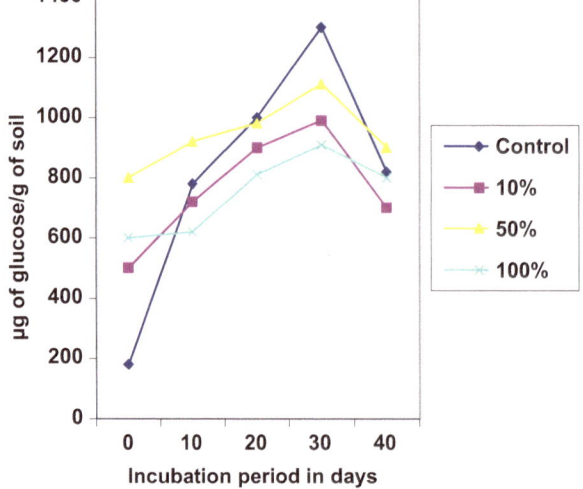

Figures in parentheses indicate relative production percentages
[a]Micrograms of glucose g^{-1} of soil
[b]Incubation, in minutes, of soil with starch (2 % w/w)

activity of amylase (72 h) increased; later at higher concentration, it declined. Similarly, by increasing the incubation period, amylase activity also increased at all concentrations, at maximum interval of 30 days; later it declined. For instance, the 10 % sample showed 500 µg GE g^{-1} activity at 0 day; it increased by 98 % at 30 days and then reduced by 19 % at 40 days interval. The same results were observed in the other three samples.

Amylase activity in soils treated with various concentrations of effluents was measured by incubating the samples for 72 h without amendment of substrate, and the results are shown in Table 7.8. By increasing the concentration of effluents, amylase activity was increased in soils up to 50 % effluent concentration; then it declined. For instance, at 0 day, 50 % effluent-treated soil exhibited 72 µg GE g^{-1}, whereas control, 10, and 100 % samples showed 20, 50, and 40 µg GE g^{-1} activity, respectively. The same trend was followed at all incubations. Another observation

Table 7.8 Amylase activity[a] in soil (without substrate) after 72 h[b] incubation as influenced by different concentrations of sugar industrial effluents

Incubation (in days)	Different concentrations of effluents (%)			
	0	10	50	100
0	20 (100)	50 (250)	72 (360)	40 (200)
10	70 (100)	150 (214)	200 (286)	102 (146)
20	92 (100)	200 (217)	300 (336)	150 (163)
30	110 (100)	250 (227)	325 (295)	220 (200)
40	25 (100)	140 (560)	150 (600)	120 (480)

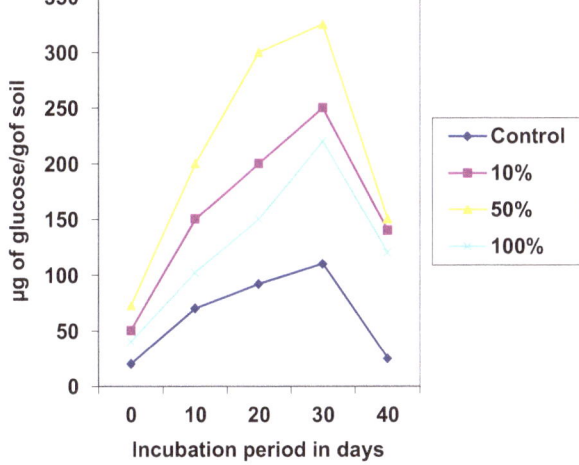

Figures in parentheses indicate relative production percentages
[a]Micrograms of glucose g^{-1} of soil
[b]Incubation, in minutes, of soil without starch (2 % w/w)

from the results was that all three samples with different concentrations of effluents showed higher amylase activities over control at all incubations. At the same time, by increasing the incubation period, amylase activity increased, at a maximum interval of 30 days, and later decreased at all concentrations. For instance, the 10 % sample showed 50 µg GE g^{-1} at 0 day, it increased by 400 % at 30 days, and then declined by 44 % at 40 days interval. The overall amylase activity without substrate was comparatively less at all incubations and concentrations with substrate.

Chapter 8
Soil Invertase

Introduction

Invertase catalyzes the hydrolysis of sucrose to glucose and fructose due to β-fructofuranosides, predominantly available in microorganisms, animals, and plants (Kiss and Peterfi 1959; Skujins 1976; Alef and Nannipieri 1995). Invertase brings out all hydrolysis of sucrose under either acidic or alkaline conditions (Splading 1979).

Very little information is available on invertase activity in soil polluted by agro-based industries. Ross and Speir (1984) reported that soil temperature influenced soil invertase activity. Invertase activity was greater in desert remnant than in xeri-scape sites (Douglas and Oleksyszyn 2002). Increased invertase activity was reported in urban expansions into wild lands (Douglas and Oleksyszyn 2002), soil treated with effluents of pulp and paper mills (Kannan and Oblisami 1990b; Chinnaiah et al. 2002), cotton ginning (Narasimha 1997), and hexachlorocyclo-hexane and its isomers (Srimathi and Karanth 1989). By contrast, invertase activity was reduced when soils were stored and air dried (Hoffman and Hoffman 1955), with the addition of toluene (Kiss and Peterfi 1959), insecticides (El Hamady and Sheloa 1999; Palaniappan and Balasubramanian 1985), cement dust from the cement industry (Shanthi 1993), and soil organic matter (Malcolm and Vaughan 1979). Bezuglova et al. (1999) reported that the higher the content of carbonates in parent rocks, the lower the activity of invertase in the buried mass, and vice versa.

Enzyme Assay

Triplicate samples of soils with/without effluent discharges were incubated in the manner mentioned in Chap. 4; soil samples were withdrawn at desired intervals at 0, 10, 20, 30, and 40 days to determine invertase activity by the method described

© Springer International Publishing Switzerland 2017   41
N.R. Maddela et al., *Soil Enzymes*, SpringerBriefs in Environmental Science,
DOI 10.1007/978-3-319-42655-6_8

by Tu (1982). Five grams of soil samples were transferred to test tubes (25 ×
200 mm); to this 1 ml of toluene was added. All the contents in the tubes were mixed
thoroughly, and after 15 min, 6 ml of 18 mmol/l sucrose in 0.2 M acetate buffer
(pH 5.5) was added. In another set, soil samples were treated in the same manner by
replacing starch with acetate buffer without substrate. All the tubes were incubated
for 6 h. The suspension was filtered by Whatman No. 1 filter paper, and the amount
of reducing sugar content in the filtrate was determined by the Nelson–Somagyi
method (1944) using an Elico digital spectrophotometer. Similarly, another three
sets of control soil samples were treated with 10, 50, and 100 % effluents, respectively, and invertase activities were assessed.

Impact of Effluents

Invertase enzyme activity was expressed as the amount of glucose formed from the
substrate (18 mM sucrose). The activity of invertase in test and control soil samples
was measured by incubating the samples in the presence of substrate (18 mM
sucrose), and the results are listed in Table 8.1. The test sample showed higher activity over control at all incubations; the increasing percentage of the test sample was
between 120 and 300 over control. Both samples showed higher activity at 30 days
interval and then activities declined. For instance, the test sample exhibited 0.48 mg
GE g^{-1} invertase activity against 0.12 mg GE g^{-1} of control at 0 day interval; later it
was increased by 766 % at 30 days and declined by 202 % at 40 days interval in test.
The increased invertase activity in polluted soil over control may be due to the availability of substrate and/or sucrose degrading microflora in polluted soil. Similar
results were reported by others: the addition of pressmud and paper mill effluent
irrigation (Chinnaiah et al. 2002), pulp and paper mill effluents (Kannan and
Oblisami 1990b), cotton ginning mill effluents (Narasimha 1997), straw under
flooded conditions (Chandrayan et al. 1980), and fungicides (Sreenivasulu 2005)
improved soil invertase activity.

The soil invertase activity in test and control soil samples were measured by
incubating without supplementation of substrate, and the results are depicted in
Table 8.2. The test sample exhibited higher activity over control at all incubations
ranging from 11 to 100. Both samples showed higher activity at 30 days interval;
later activities were lower. For instance, the test sample exhibited 0.06 mg GE g^{-1}
activity at 0 day against 0.03 mg GE g^{-1} of control. It increased by 416 % at 30 days
and declined by 68 % at 40 days in test. Comparatively at all incubations, the invertase activity without substrate in test and control was less with substrate.

Invertase activity in soils treated with different concentrations of effluents such
as 10, 50, and 100 % was measured by incubating with the addition of substrate
(18 mM sucrose), were described, and the results reported in Table 8.3. By increasing the concentrations of the effluents, invertase activity was increased up to 10 %;

Table 8.1 Invertase activity[a] in soil (with substrate) after 6 h[b] incubation as influenced by sugar industrial effluents

Incubation (in days)	Control[c]	Test[d]
0	0.12 (100)	0.48 (400)
10	0.48 (100)	1.82 (379)
20	1.33 (100)	3.10 (233)
30	1.89 (100)	4.16 (220)
40	1.57 (100)	2.40 (136)

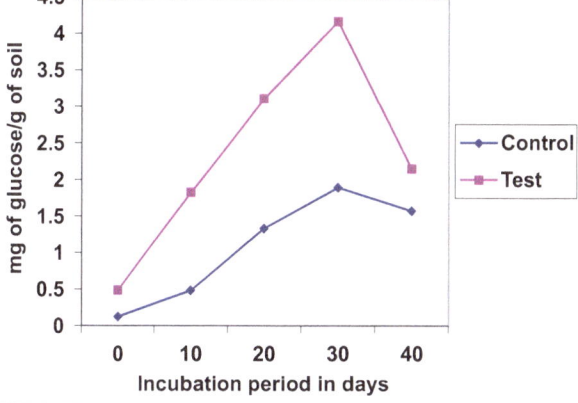

Figures in parentheses indicate relative production percentages
[a]Milligrams of glucose g^{-1} of soil
[b]Incubation, in hours, of soil with sucrose (18 mM)
[c]Soil without sugar industrial effluents
[d]Soil polluted with sugar industrial effluents

later it decreased in soils. At lower concentrations of effluents (10 %), samples showed higher activity and vice versa. For instance, at 0 day, 10 % effluent-treated soil exhibited 0.4 mg GE g^{-1}, whereas control, 50, and 100 % samples showed 0.12, 0.26, and 0.16 mg GE g^{-1} invertase activity, respectively. The same trend was followed at all incubations. Similarly, by increasing the incubation period, invertase activity also increased at all concentrations, at maximum interval of 30 days; later it declined. For instance, the 10 % sample showed 0.4 mg GE g^{-1} activity at 0 day, it increased by 507 % at 30 days, and then reduced by 43 % at 40 days interval. The same results were observed in the remaining three samples.

Invertase activity in soils treated with various concentrations of effluents was measured without amendment of substrate, described in the section "Enzyme Assay," and the results are reported in Table 8.4. Here also similar results were obtained. By increasing the concentration of effluent, the invertase activity increased up to 10 % effluent concentration and later it decreased. For instance, at 0 day interval, the 10 % effluent-treated sample showed 0.05 mg GE g^{-1} activity,

Table 8.2 Invertase activity[a] in soil (without substrate) after 6 h[b] incubation as influenced by sugar industrial effluents

Incubation (in days)	Control[c]	Test[d]
0	0.03 (100)	0.06 (200)
10	0.06 (100)	0.12 (200)
20	0.18 (100)	0.22 (122)
30	0.28 (100)	0.31 (111)
40	0.25 (100)	0.10 (40)

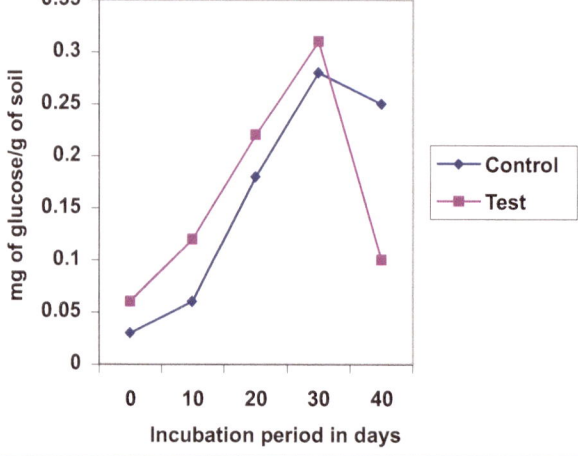

Figures in parentheses indicate relative production percentages
[a]Milligrams of glucose g^{-1} of soil
[b]Incubation, in hours, of soil without sucrose (18 mM)
[c]Soil without sugar industrial effluents
[d]Soil polluted with sugar industrial effluents

whereas the control, 50, and 100 % samples exhibited 0.03, 0.03, and 0.01 mg GE g^{-1} activities, respectively. At the same time, by increasing the incubation period, the invertase activity also increased, at a maximum interval of 30 days interval, and later decreased in all samples. For instance, the 10 % sample showed 0.05 mg GE g^{-1} at 0 day, it increased by 380 % at 30 days, and then declined by 26 % at 40 days interval. The overall invertase activity without substrate was comparatively less at all incubations and concentrations with substrate.

Table 8.3 Invertase activity[a] in soil (with substrate) after 6 h[b] incubation as influenced by different concentrations of sugar industrial effluents

Incubation (in days)	Different concentrations of effluents (%)			
	0	10	50	100
0	0.12 (100)	0.40 (333)	0.26 (217)	0.16 (133)
10	0.48 (100)	0.82 (171)	0.50 (104)	0.41 (85)
20	1.33 (100)	1.5 (113)	1.2 (90)	0.9 (68)
30	1.89 (100)	2.43 (128)	2.20 (116)	1.70 (90)
40	1.57 (100)	1.40 (89)	1.3 (83)	0.52 (33)

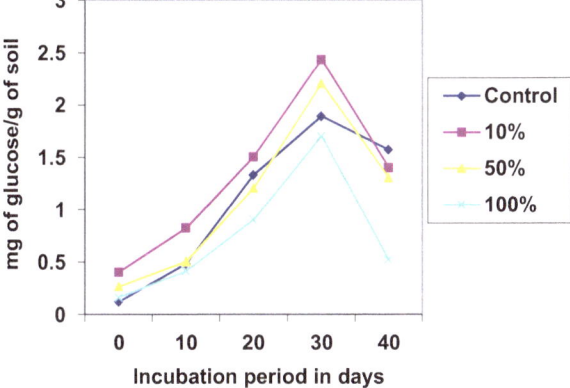

Figures in parentheses indicate relative production percentages
[a]Milligrams of glucose g^{-1} of soil
[b]Incubation, in hours, of soil with sucrose (18 mM)

Table 8.4 Invertase activity[a] in soil (without substrate) after 6 h[b] incubation as influenced by different concentrations of sugar industrial effluents

Incubation (in days)	Different concentrations of effluents (%)			
	0	10	50	100
0	0.03 (100)	0.05 (167)	0.03 (100)	0.01 (33)
10	0.06 (100)	0.08 (133)	0.04 (67)	0.02 (33)
20	0.18 (100)	0.10 (55)	0.06 (33)	0.06 (33)
30	0.28 (100)	0.19 (68)	0.08 (28)	0.07 (25)
40	0.25 (100)	0.14 (96)	0.05 (20)	0.026 (100)

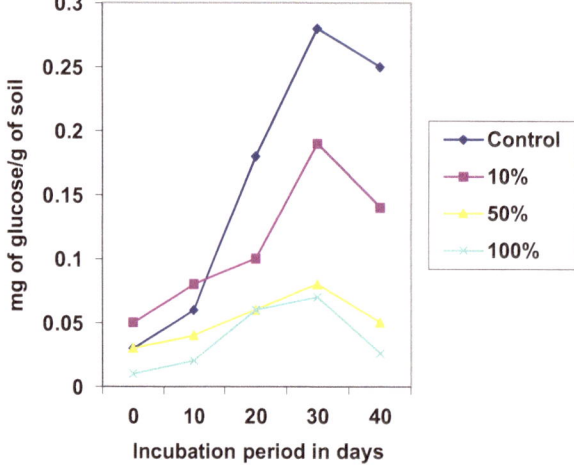

Figures in parentheses indicate relative production percentages
[a]Milligrams of glucose g^{-1} of soil
[b]Incubation, in hours, of soil without sucrose (18 mM)

References

Abdelnainm EM, Rao MS, Wally TM, Nashar EMB (1987) Effect of prolonged sewage irrigation on some physical properties of sandy soil. Bio Wastes 22:269–274

Achberger K, Ohlinger R (1988) Effect of sewage sludge and waste compost on some enzymatic activities tested in a field experiments. Poster presentation at EC. EWPCA symposium, Amsterdam

Adhikari S, Mitra A, Gupta SK, Banarjee SK (1994) Proc Indian Nat Sci Academy, Part-B Biol. Sciences 60(6):541–552

Ahn MY, Jerzy D, Jang-Eok K, Jean-Marc B (2002) Treatment of 2,4-dichlorophenol polluted soil with free and immobilized laccase. J Environ Qual 31:1509–1515

Alef LK, Nannipieri P (1995) Methods in soil microbiology and enzyme activities. Academic Press/Harcourt Brace and Company Publishers, London, pp 225–230

Alexander M (1961) Introduction to soil microbiology. Wiley Eastern Ltd, New Delhi

Alexander M (1977) Introduction to soil microbiology, 2nd edn. Wiley, New York

Amann RI, Ludwig W, Schleifer KH (1995) Phylogenetic identification and in situ detection of individual microbial cells without cultivation. Microbiol Rev 59:143–169

Andrade ML (2002) Industrial impact of marsh soils at the Bahia Blanca Ria, Argentina. J Environ Qual 31:532–538

Anikew MAN (2002) Long term effect of municipal waste disposal on soil properties and productivity of sites used for urban agriculture in Abakaliki, Nigeria. Info Bioresour Technol 83(3):241–250

Bezuglova OS, Gorbov SN, Evseeva NV, Urkova EG (1999) Changing of the urban soils biological activity in the case of sealing-up. Enzymes in the environment: activity, ecology, applications. Granada, Spain. July, 12–15, pp 122

Bhogal NS, Prasad P, Sakal R (2002) Phyto accumulation of micronutrients and pollutants in calciorthent receiving sewage effluents in India. Poster presentation, paper no. 1949, symposium no. 24. 17th WCSS. Biol Biochem 31:1471–1479

Bollag JM, Berthelin J, Adriano D, Huang PM (2002) Impact of soil minerals-organic component — microorganisms interactions on restoration of terrestrial ecosystems. Poster presentation, paper no. 861, symposium no. 47. 17th WCSS

Brady N, Weil R (2002) The nature and properties of soils, 13th edn. Prentice Hall, Upper Saddle River, NJ, 960 pp

Chandrayan K, Adhya TK, Sethunathan N (1980) Dehydrogenase and invertase activity of flooded soils. Soil Biol Biochem 2:127–137

Chinnaiah U, Palaniappan M, Augustine S (2002) Rehabilitation of paper mill effluent polluted soil habitat: an Indian experience. Poster presentation, paper no. 770, symposium no. 24. 17th WCSS

© Springer International Publishing Switzerland 2017

N.R. Maddela et al., *Soil Enzymes*, SpringerBriefs in Environmental Science, DOI 10.1007/978-3-319-42655-6

Chuasavathi T, Trelo-ges V (2001) An improvement of Yasothon soil fertility (Oxic Palanstults) using municipal fermented organic compost and *Panicum maximum* TD 58 grass. Pak J Biol Sci 4(8):968–972

Cole MA (1977) Lead inhibition of enzyme synthesis in soil. Appl Environ Microbiol 33:262–268

Craul PJ (1992) Urban sol in landscape design. Wiley, New York

Dedeken M, Voets JP (1965) Research on the metabolism of amino acids in the soil. I. The metabolism of glycine, alanine, aspartic acid and glutaminic acid. Ann Inst Pasteur (Paris) 3:103–111

Devarajan L, Satisha GC, Nagendran K (2002) Distillery effluent—a source for fertilization and composting of pressmud and other biodegradables. Poster presentation, paper no. 891, symposium no. 24. 17th WCSS

Devi S, Datta A, Surya Rao P (2002) Evaluation of maturity of compost based on coir dust: an agro-industrial waste in India. Poster presentation, paper no. 715, symposium no. 24. 17th WCSS

Donahue RL, Miller RW, Shickluna JC (1983) Soils—an introduction to soils and plant growth, 5th edn. Prentice-Hall, Englewood Cliffs, NJ

Donnelly PK, Entry JA, Crawford DL, Cromack K Jr (1990) Cellulase and lignin degradation in forest soils response to moisture, temperature and acidity. Microb Ecol 20:289–295

Douglas MG, Oleksyszyn M (2002) Enzyme activities and carbon dioxide flux in a Sonoran desert urban ecosystem. Soil Sci Soc Am J 66:2002–2008

El Hamady SEE, Sheloa MKAA (1999) Field evaluation of imidacloprid applied as seed treatment to control thrips tabaci lind on cotton with regard to soil pollution. Arab Univ J Agric Sci 7(2):561–574

Frey SD, Elliot ET, Paustian K (1999) Bacterial and fungal abundance and biomass in conventional and no-tillage agroecosystems along two climatic gradients. Soil Biol Biochem 31(4): 573–585

Gaianfreda L, Bollag JM (1994) Effect of soils on the behavior of immobilized enzymes. Soil Sci Soc Am J 58:1672–1681

Gaianfreda L, Bollag JM (1996) Influence of natural and anthropgenic factors on enzyme activity in soil. In: G. Stotzky and J.M. Bollag (ed.) Soil biochemistry. Vol. 9. Marcel Dekker, New York, pp 123–193

Gardiner DT, Miller RW (2004) Soils in our environment, 10th edn. Pearson Education Inc., Upper Saddle River, NJ, 641 pp

Gilbert OL (1991) The ecology of urban habitats. Chapman and Hall, New York

Handa SK, Agnihothri MP, Kulshresta G (2000) Effect of pesticides on soil fertility. In: Pesticide residue analysis and significance. Research Periodicals and Publishing House, New Delhi, pp 184–198

Harris JA (1991) The biology of soils in urban areas. In: Bullock P, Grgory PJ (eds) Soils in the urban environment. Blackwell Scientific Publications, Oxford, pp 139–152

Hayano K (1986) Cellulase complex in tomato field soil; introduction localization and some properties. Soil Biol Biochem 18:215–219

Hoffman E, Hoffman G (1955) About the enzyme system of our culture ground, VI. Amylase. J Plant Nutr Soil Sci 70:97–104

Jackson ML (1971) Soil chemical analysis. Prentice Hall, New Delhi

Johnson CM, Ulrich A (1960) Determination of moisture in plant tissues. Calif Agri Bull, No. 766. In: Wilde SA et al (ed) Soil and plant analysis for tree culture. Obortage Publishing Co., Oxford/Bombay, pp 112–115

Joshi SR, Sharma GD, Mishra RR (1993) Microbial enzyme activities related to litter decomposition near a highway in a sub tropical forest of North East India. Soil Biol Biochem 22:51–55

Kanazawa S, Miyashita K (1987) Cellulase activity in forest soil. Soil Sci Plant Nutr 33:399–406

Kannan K, Oblisami G (1990a) Influence of irrigation with pulp and paper mill effluent on soil chemical and microbiological properties. Biol Fertil Soils 10:197–201

Kannan K, Oblisami G (1990b) Influence of pulp and paper mill effluents on soil enzyme activities. Soil Biol Biochem 22:923–927

Kim YW, Kim KY, Lee JJ, Shim JH, Park RD, Kim KS, Sohn BK, Chung SJ (2002) Effect of food waste compost on microbial population, enzyme activity, and lettuce growth. Poster presentation, paper no. 921, symposium no. 24. 17th WCSS

Kiss S, Peterfi Jr (1959) Biologia. 2:179 (cited in C.M. Tu, 1982, Chemosphere, 11:909–914)

Kong KT, Dommergues (1972) Limitation ele cellulose dansles sols organiques, 11. Elludae des enzymes du Sol Rev Ecol Biol Sol 9:629–640

Ladd JN, Butler JHA (1969) Inhibitory effect of soil humic compounds on the proteolytic enzyme, pronase. Aust J Soil Res 7:241–251

Ladd JN, Butler JHA (1972) Short-term assays of soil proteolytic enzyme activities using proteins and dipeptide derivatives as substrates. Soil Biol Biochem 4:19–30

Lindsay WL (1979) Chemical equilibria in soils. Wiley, New York, 449 pp

Linkins AE, Mellio JM, Sinsabaugh RL (1984) Factors affecting cellulase activity in terrestrial and aquatic ecosystems. In: Klug MJ, Reddy CA (eds), Current perspectives in microbial ecology, American Society for Microbiology, Washington, DC, pp 572–579

Linkins AE, Sinsabaugh RL, Mc Clargherty CA, Mellilo JM (1990) Cellulase activity on decomposing leaf litter in microcosms. Plant and Soil 123:17–25

Lowry OH, Rosebrough NJ, Farr AL, Randall RJ (1951) Protein measurement with the Folin phenol reagent. J Biol Chem 193(1):265–275

Maile WH, Linkins AE (1978) Cellulase activity during the growth of *Achlya bisexualis* on glucose, cellulose and selected polysaccharides. Can J Bot 56:1974–1981

Malcolm RE, Vaughan D (1979) Effect of humic acid fraction on invertase activities in plant tissues. Soil Biol Biochem 11:65–72

Mandic L, Dukic D, Govedarica M, Stamenkovic S (1997) The effect of some insecticides on the number of amylolytic microorganisms and azotobacter in apple nursery soil. Czeckoslevensko Vocarstvo 31(1–2):177–184

Markus L (1955) Determination of carbohydrates from plant materials with Anthrone reagent assay of cellulase activity in soil and farmyard manure. Agrokem Talajt 4:207–216

Massoud F (1972) Some physical properties of highly calcareous soils and their related management practices. FAO/UNDP regional seminar on reclamation and management of calcareous soils. Cairo, Egypt. Nov 27–Dec 2, 1972. http://www.fao.org/docrep/x5868e/x5868e00.-htm#Contents. Accessed June 2004

Medhi UJ, Talukdar AK, Deka S (2005) Physicochemical characteristics of lime sludge waste of paper mill and its impact on growth and production of rice. J Ind Pollut Control 21(1):51–58

Mishra PC, Sunandashaoo (1989) Agropotentiality of paper mill waste water. In: Soil pollution and soil organism. (PC Mishra Eds), Ashish Publishing House, New Delhi, pp 97–119

Monanmani K, Chitraraju G, Swaminathan K (1990) Effect of alcohol and chemical industrial effluents on physical and biological properties of soil. Poll Res 9:79–82

Montuelle B, Volat B (1998) Impact of wastewater treatment plant discharge on enzyme activity in freshwater sediments. Ecotoxicol Environ Saf 40(1–2):154–159

Moreno JL, Garcia C, Hernandez T (2003) Toxic effect of cadmium and nickel on soil enzymes and the influence of adding sewage sludge. Eur J Soil Sci 54:377–386

Nandakumar NV (1990) Tannary and chromate industries effluents effect on soil, animals and plants. In: Mishra PC (ed) Soil pollution and soil organisms. Ashish Publishing House, New Delhi, pp 81–105

Narasimha G (1997) Effect of effluent of cotton ginning industry soil microbial activities. M. Phil., Dissertation submitted to Sri Krishnadevaraya University, Anantapur

Narasimha G, Babu GVAK, Rajasekhar Reddy B (1999) Physico-chemical and biological properties of soil samples collected from soil contaminated with effluents of cotton ginning industry. J Environ Biol 20:235–239

Nelson N (1944) A photometric adaptation of Somogyi method for determination of glucose. J Biol Chem 153:375–380

Nichols KA, Wright SF, Liebig MA, Pikul JL Jr (2004) Functional significance of glomalin to soil fertility. Proceedings from the Great Plains soil fertility conference proceedings, Denver, CO, 2–4 Mar 2004

Olfert O, Johnson GD, Brandt SA, Thomas AG (2002) Use of arthropod diversity and abundance to evaluate cropping systems. Agron J 94:210–216

Omar SA, Abd-Alla MA (2000) Microbial populations and enzyme activities in soil treated with pesticides. Water Air Soil Pollut 127(1–4):49–63

Pahwa SH, Bajaj K (1999) Effect of pre-emergence herbicides on the activity of α-amylase and protease enzyme during germination in pigeon pea and carpet weed. Indian J Weed Sci 31(2–4):148–150

Palaniappan SP, Balasubramanian A (1985) Influence of two pesticides on certain soil enzymes. Agric Res J Kerala 23(2):189–192

Pancholy SK, Rice EL (1973) Soil enzymes in relation to old field succession: amylase, cellulase, invertase, dehydrogenase and urease. Soil Sci Soc Am Proc 37:47–50

Plaza C, Garcia-Gil JC, Soler-Revira P, Polo A (2002) Effect of agricultural application of pig slurry on soil enzyme activities. Poster presentation. Centro de Ciencias Medioambientales (CSIC), Madrid, Espana

Ramakrishna Parama VR, Venkatesha M, Bhargavi MV (2002) Recycling of urban domestic residues as a nutrient source for agriculture. Poster presentation, paper no. 904, symposium no. 24. 17th WCSS

Rawald LW, Domke K, Stohr G (1968) Slides on the relations between humus quality and microflora of soil. Pedobiologia 7:375–380

Renukaprasanna M, Channal HT, Sarangamath PA (2002) Characterization of city sewage and its impact on soils and water bodies. Poster presentation, paper no. 70, symposium no. 24. 17th WCSS

Richards BN (1987) The microbiology of terrestrial ecosystems. Longman Scientific and Technical, Essex

Ross DJ, Roberts HS (1973) Biochemical activities in soil profile under hard beech forest. Invertase, and amylase activities and relationships, with other properties. N Z J Sci 16:209–224

Ross DJ, Speir TW (1984) Temporal fluctuations in biochemical properties of soil under pasture. II. Nitrogen mineralization and enzyme activities. Aust J Soil Res 22:319–330

Ruggiero P, Dec J, Bollag JM (1996) Soil as a catalytic system p. In: Stotzky G, Bollag JM (eds) Soil biochemistry, vol 9. Marcel Dekker, New York, pp 79–122

Sarade R, Richard J (1994) Characterization and enumeration of micro-organisms associated with anaerobic digestion of tomato processing waste. Bioresour Technol 49(3):261–265

Shanthi M (1993) Soil biochemical processing industrially polluted areas of cement industry. M. Phil., Dissertation, Sri Krishnadevaraya University, Anantapur

Singaram P, Kamalakumari K (2000) Effect of continuous application of different levels of fertilizers and farm yard manure on enzyme dynamics of soil. Mad Agric J 87(4, 6):364–365

Singh BK, Allan W, Denis JW (2002) Degradation of chlorpyrifos, fenamiphos, and chlorothalonil alone and in combination and their effects on soil microbial activity. Environ Toxicol Chem 21(12):2600–2605

Singh SP, Bhutnagar MK, Pritishrivasstava, Ablilungha singh (2005) Growth performance and chemical analysis of some plants irrigated with paper mill effluents. J Ind Pollut Control 21(1):163–166

Sinsabaugh RL, Linkins AE (1987) Inhibition of the *Trichoderma viridae* cellulase complex by leaf litter extracts. Soil Biol Biochem 19:719–725

Sivakumar S, De Brito JA (1995) Effect of cement pollution soil fertility. J Ecotoxico Environ Monit 5(2):147–149

Skujins J (1976) Extracellular enzymes in soil. CRC Crit Rev Microbiol 4:383–421

Smith SR (1991) Effect of sewage of sludge application on soil microbial processes and soil fertility. Adv Soil sci 16:191–203

Smith SE, Read DJ (1997) Mycorrhizal symbiosis, 2nd edn. Academic Press, San Diego, CA, 605 pp

Sparkling GP, Cheshire MV (1979) Effects of soil drying and storage on subsequent microbial growth. Soil Biol Biochem II:317–319

Speir TW, Ross DJ (1975) Effects of storage on the activities of protease, urease, phosphatase and sulphatase in three soils under pasture. N Z J Sci 18:231–237

Splading BP (1979) Effect of divalent metal cations respiration and extractable enzymes activities of Douglss-fir needle litter. J Environ Qual 8:105–109

Sreenivasulu M (2005) Interactions between tridemorph and captan (Fungicides) with microorganisms in ground nut (*Arachis hypogaea* L.) Soils. M. Phil., Dissertation submitted to Sri Krishnadevaraya University, Anantapur

Srimathi MS, Karanth MGK (1989) Influence of hexachlorocyclohexane isomers on soil enzymes. J Soil Biol Ecol 9(2):65–71

Stotzky G, Goos RD, Timonin MI (1962) Microbial changes and occurring in soil as a result of storage. Plant Soil 16:1–18

Swaminathan K, Ravi K (1987) Effect of dying factory effluents on physico-chemical and biological properties of soil. In: Dalela RC, Sahai YN, Gupta S (eds) Environment and ecotoxicology. The Academy of Environmental Biology, Muzaffarnagar, pp 249–253

Tu CM (1981a) Effect of pesticides on activity of enzymes and microorganisms in a clay loam soil. J Environ Sci Health 16:179–181

Tu CM (1981b) Effect of some pesticides on enzyme activities in an organic soil. Bull Environ Contam Toxicol 27:109–114

Tu CM (1982) Influence of pesticides on activities of amylase, invertase and level of adenosine triphosphate inorganic soil. Chemosphere 2:909–914

Tu CM (1995) Effect of five insecticides on microbial and enzyme activities in sandy soil. J Environ Sci Health 30(3):289–306

Tu CM, Miles JRW (1976) Interactions between insecticides and soil microbes. Res Rev 64:17–65

Tugel AJ, Lewandowski AM (eds) (1999) Soil biology primer. NRCS Soil Quality Institute, Ames, IA, 50 pp

Vigil MF, Sparks D (2003) Conservation tillage fact sheet. Central Great Plains research station, Akron, CO. http://www.akron.ars.usda.gov/fs_factors.html. Accessed Feb 25, 2004

Walker JD, Austin HF, Colwell RR (1975) Utilization of mixed hydrocarbon substrate by petroleum-degrading microorganisms. J Gen Appl Microbiol 21:27–39

Wollum AG (1994) Soil sampling for microbiological analysis. In: Methods of soil analysis, part 2 — microbiological and bio-chemical properties. SSSA Book Series No. 5. Soil Science Society of America, Madison, WI, pp 2–13

Wood TM, Bat KM (1988) Methods for measuring cellulase activities. In: Wood W, Kellogg SJ (eds) Methods in enzymology, vol 160. Academic, New York, p 112

Xiao C, Fauci M, Bezdicek DF, McKean WT, Pan WL (2005) Soil microbial responses to potassium-based black liquor fro straw pulping. Soil Sci Soc Am J 70:72–77

Zende GK (1995) Sugar industry by-products and crop residues in increasing soil fertility and crop productivity. In: Singh GB, Solomon S (eds) Sugarcane – agro industrial alternatives. Oxford IBH, India, pp 351–370